Piracy of Digital Content

OECD

ORGANISATION FOR ECONOMIC CO-OPERATION AND DEVELOPMENT

The OECD is a unique forum where the governments of 30 democracies work together to address the economic, social and environmental challenges of globalisation. The OECD is also at the forefront of efforts to understand and to help governments respond to new developments and concerns, such as corporate governance, the information economy and the challenges of an ageing population. The Organisation provides a setting where governments can compare policy experiences, seek answers to common problems, identify good practice and work to co-ordinate domestic and international policies.

The OECD member countries are: Australia, Austria, Belgium, Canada, the Czech Republic, Denmark, Finland, France, Germany, Greece, Hungary, Iceland, Ireland, Italy, Japan, Korea, Luxembourg, Mexico, the Netherlands, New Zealand, Norway, Poland, Portugal, the Slovak Republic, Spain, Sweden, Switzerland, Turkey, the United Kingdom and the United States. The Commission of the European Communities takes part in the work of the OECD.

OECD Publishing disseminates widely the results of the Organisation's statistics gathering and research on economic, social and environmental issues, as well as the conventions, guidelines and standards agreed by its members.

This work is published on the responsibility of the Secretary-General of the OECD. The opinions expressed and arguments employed herein do not necessarily reflect the official views of the Organisation or of the governments of its member countries.

Corrigenda to OECD publications may be found on line at: *www.oecd.org/publishing/corrigenda*.

Foreword

This study covers the piracy of content, such as music, films and software that does not involve the use of physical media; in other words it essentially deals with piracy over the Internet. This represents Phase II of a three-phase project examining the Economic Impacts of Counterfeiting and Piracy *(www.oecd.org/sti/counterfeiting)*. Phase I covered counterfeiting and piracy resulting in the production of tangible goods, while Phase III will cover other forms of counterfeiting.

The study was prepared by the Structural Policy Division of the OECD Directorate for Science, Technology and Industry. The main study was prepared by Piotr Stryszowski, while the case study was prepared by Danny Scorpecci. The study was carried out under the guidance of an informal expert advisory group composed of OECD member governments and industry representatives, and the study team wishes to acknowledge the valuable assistance provided by the advisory group.

The study was presented to the OECD Committee on Industry, Innovation and Entrepreneurship, which agreed to its public release. This study was also presented to the OECD Information, Computer and Communications Policy Committee and benefited from this committee's comments. The study was undertaken with the support of governments and industry associations, and the authors express their gratitude for the inputs provided.

Table of Contents

Executive Summary

This study of digital piracy covers the infringement of copyrighted content (such as music, films, software, broadcasting, books, etc.) where the end product does not involve the use of hard media, such as CDs and DVDs. Piracy involving hard media was covered in Phase I of the overall study.[1]

The study highlights the unique market dynamics of digital piracy,[2] where the existence of a large number of suppliers willing to provide pirated content at virtually zero price pose new and difficult challenges to copyright owners and policy makers in combating that piracy.

What is digital piracy?

While there are definitions of "pirated copyright goods", there is currently no specific legal definition of digital piracy, which would be more accurately described as "digital infringement of copyright".

This lack of a common definition is reflected in the different treatment of copyright infringement in different jurisdictions, which in varying degrees provide exceptions to the rights of the copyright owner, such as fair dealing, fair use and private and domestic use. What is illegal or criminal in one jurisdiction might not always be illegal or criminal in another.

Digital market characteristics

Suppliers and consumers are sometimes one and the same

Even though technically the markets for pirated digital products are made up of suppliers and consumers, in many cases a single individual serves as both. For example, this can happen when suppliers expect future reciprocity, or in some peer-to-peer networks (such as Bit Torrents) where consumers can be simultaneous suppliers of digital content.

In the digital piracy market the profit motive can be absent

This market is further complicated by the existence of a large number of suppliers whose principal objective is not profit, but who are driven by other non-market factors, such as gaining recognition within a peer group, or reciprocating free access to other users. This behaviour can be sustained because the marginal cost of reproduction and delivery of digital content is zero, or close to zero.

This in turn has created unique market dynamics…

Unlike other markets, where owners of copyright affected by piracy have to contend with pirates who will supply similar content at a much lower price, in this market they also need to compete with suppliers willing to provide content at zero price. In this context, non-price factors (such as legality, availability and quality) become especially important in allowing suppliers that require payment to continue to operate in the market.

While the rationale for the existence of an illegal sub-market where pirated products are exchanged against a price seems counterintuitive (in the face of competition from legitimate providers and free pirates sites), they nevertheless do exist, probably by servicing niche markets. When they do operate, they are likely to enjoy exceptionally high profits by exploiting their low cost operations, especially if they are able to mislead users into thinking that they are legitimate, and can price their services accordingly.

…with significant consequence for copyright owners and policy makers

A by-product of this new market dynamic is that the number of suppliers of pirated digital content (many of whom do not consider themselves as "pirates") has exploded, making detection and response much more difficult and costly for copyright owners and law enforcers. This is compounded by the fact that "pirates" operate globally in different jurisdictions with different laws and regulations, which hampers the efficiency of enforcement and makes it more difficult and costly. Economies with strong copyright protection tend to report lower rates of piracy, but the risk of penalties without effective enforcement does not always seem to succeed as a strong deterrent. Moreover, the flexibility of digital piracy allows pirates to easily shift their activities to markets where legal regimes are weaker.

In addition, the large number of very diffuse suppliers and customers means that it is also difficult to increase the awareness of offenders of the negative short and long-term effects of their actions.

What factors shape digital piracy?

The low cost of reproduction and delivery is a major factor

Digital products are very much in demand, especially by younger people, and the ease and very low cost of reproduction and transmission of digital products makes these very attractive to share.

Those involved in the digital infringement of copyright also take advantage of a characteristic of this market where users often place convenience above the ultimate quality of the digital product, especially when immediacy is a factor (such as film premieres or the live broadcast of a sporting event).

Technology is a great facilitator

Computing power and the Internet facilitate the collection and transmission of digital content, and this lubricates exchanges between suppliers and customers. The global nature of the Internet means that the market is potentially huge.

In turn, this has spawned on-line services aimed at facilitating data exchanges over the Internet. For example peer-to-peer networks (P2P) facilitate the sharing of digital content, and most P2P networks encourage users to be both consumers and suppliers, as this increases the efficiency of the network. In these networks heavy uploaders are usually rewarded with better service. One-click and Warez sites also facilitate these exchanges, but without the need or opportunity to simultaneously be a supplier as well as a customer. However, it should also be recognised that not all of these services were created for the purpose of facilitating piracy, and it should not be presumed that all users of these services infringe copyright.

The recent increases in storage capacities, including web servers, local hard drives and other portable storage facilities are also factors driving both the supply and demand for pirated digital content, and the reduction in the cost of storage media can also act as a driver of digital piracy.

Source material is easy to find

Sources of digital material are easy to find, and can extend from local sources (such as family, fellow students or colleagues) to the Internet, and the ability of digital content to be transmitted with very little loss of quality means that even high quality source material can be easily obtained.

Live broadcasts (such as sporting events) can also be captured by simple video cards installed in computers, and then streamed virtually in real-time by using Unicast sites (streaming individually to end users) or P2P networks by joining a "swarm" with others interested in the content.

Digital piracy is not often perceived as unethical

Existing studies suggest that consumers of pirated digital products are mostly aware that digital piracy is illegal. However, the exchange of pirated digital products free of charge is rarely perceived as unethical, given consumers' perception that no monetary profits are generated by the infringing parties from users.

In addition, many engaged in digital piracy are attracted to the possibility of being recognised within their peer groups, or for their technical capabilities, and it is status, rather than possible financial reward, that drives their activities. Collectivism within a group, and the desire to reciprocate for received digital content, also act as drivers. Lack of parental supervision was also identified as a factor contributing to the intensity in which younger individuals engage in digital piracy.

Security problems are not fully taken into account

Users that engage in digital piracy generally appear to be unaware of potential security risks associated with their file exchanging activities, which could leave them open to exposure to malicious software which is designed to infiltrate or damage a computer system. Even those that have some knowledge of potential security threats do not appear to consider these risks as significant.

Institutional and industry responses

There are often no specific remedies for digital piracy

In general, national laws and regulations do not distinguish between digital and non-digital uses of copyrighted materials.[3] As such, available legal remedies, be they civil proceedings or action under public law, are those that exist for infringements of copyright more generally.

These remedies can sometimes not be appropriate; for example, in the instance of live broadcasting, the value of the product is greatest while it is being broadcast, while legal or technical remedies may not be able to be brought to bear in that short window of opportunity, which means that the initial loss cannot be prevented.

The global and ephemeral nature of digital piracy makes it difficult to combat

Unlike counterfeiting and piracy that results in physical goods, as a practical matter digital piracy generally cannot be detected at national borders. The flow of pirated digital products is more difficult to track by law enforcement agencies from sellers, via distributors, to producers, than the flow of physical goods. The large numbers of individuals involved, and the often observed absence of a monetary transaction, present further challenges for effective international co-operation amongst enforcement agencies. However, governments and law enforcement agencies worldwide have some successful examples of co-operation in multi-jurisdictional investigations and prosecutions of other Internet-related conduct.[4] There are likely to be lessons learned from those efforts.

Issues for policy makers and industry to consider

The digital piracy market is very different from that of counterfeit and pirated physical goods. This creates special and unique problems for copyright owners and can multiply problems faced by policy makers and producers of digital products.

Existing laws and regulations may be too broad and general to deal adequately with the rapid technological developments that facilitate digital piracy, and policy makers may need to consider enacting some specific provisions to deal with these infringements.

Such provisions should not unduly impede legitimate digital communications, nor unreasonably impact on the Internet as an effective communications platform, commercial channel and educational tool – see the recommendation of the June 2008 Seoul Ministerial Declaration on the Future of the Internet Economy *(www.oecd.org/futureinternet)*, which provides important context in this regard.

Where digital piracy affects copyright products of a highly perishable nature (such as the live broadcasting of sports events, pre-release or simultaneous release of creative content), policy makers may need to consider ways of improving the speed at which the legal system responds to these infringements.

Legal action against web sites that sustain or promote the illegal transfer of digital content and action against individuals infringing copyright could still bring some results, and still remain an important part of the overall approach to deal with Internet piracy. However, the overall efficiency of such actions will be diminished over time, because of the potentially very large number of actors involved in different jurisdictions, and who operate in a decentralised manner independent of any central hub. Thus, new ways of dealing with digital piracy may be required including new legal approaches and education. These new ways need to be fair and equitable for all parties and in particular, must reflect a balance between the interests of copyright owners, users and intermediaries.

Even if consumers involved in digital piracy are generally aware that it is illegal, it is not always perceived as un-ethical, so continued emphasis should be placed on education and consumer awareness to overcome this perception. Industry and some governments have instituted extensive campaigns toward this end, and these actions can be paralleled with enforcement campaigns to reduce the likelihood of piracy. Policy makers should consider the role that public education and consumer awareness outreach plays to help users understand what is legal and illegal. Industry and some governments have instituted extensive campaigns towards this end. The global nature of digital piracy means that international co operation between governments, enforcement agencies, industry and consumers is even more important (but more difficult) than in other sectors, especially in identifying and taking down web sites that encourage or promote the transfer of pirated digital content.

In some market segments, business models are already evolving to respond to new market dynamics brought about by the low cost of reproduction and distribution of digital content. Consumers, who are offered attractive legal options, have less incentive to use the illegal alternatives. In this context, factors such as legality, quality and ease of use are becoming key factors in attracting customers. However, the pervasiveness of and easy access to unauthorised content operate as significant limiting factors in efforts to fully develop a robust, legitimate marketplace. Moreover, the development of new business models and content services that meet consumers' expectations should not be overlooked when considering ways to address piracy.

Notes

1. See *www.oecd.org/sti/counterfeiting*.

2. In the context of this study, the term "digital piracy" (as opposed to "physical piracy") refers to piracy that does not involve the use of hard media.

3. There are some significant exceptions, such as the US "No Electronic Theft" Act (NET).

4. The OECD Recommendation on Cross-Border Co-operation in Enforcement of Laws against Spam, and the OECD Recommendation on Cross-Border Co-operation in the Laws Protecting Privacy, are examples.
   See *www.oecd.org/sti/cultureofsecurity*.

Introduction

In response to rising concerns in government and the business community, the OECD has launched a project to assess the economic effects of counterfeiting and piracy. The objective of the project is to improve factual understanding and awareness of the effects that infringements of intellectual property rights, as described and defined in the World Trade Organization (WTO) Agreement on Trade-Related Aspects of Intellectual Property Rights (TRIPS), have on governments, business and consumers in member countries and non-member economies.

The problem with counterfeiting and piracy is seen by governments, industry and affected labour groups as growing in complexity, and as well as the economic impacts there are often potentially serious health and safety consequences associated with an expansion in the quantity and range of fake products being marketed internationally. There is also increasing concern that criminal networks (including organised crime and terrorist groups) are among the principal beneficiaries of counterfeiting and piracy activities, with the proceeds from their activities being used to finance a range of illicit activities.

The project is being carried out in three separate phases. The Phase I report (OECD, 2008) published in June 2008 focuses on tangible products that infringe trademarks, copyrights, patents and design rights. Phase II partially covers digital piracy and Phase III will deal with all other infringements of intellectual property rights (IPR).

The objective of Phase II of the project on the Economic Impacts of Counterfeiting and Piracy is to assess the impact of piracy of digital content. The OECD Council mandate used the term "digital piracy" to mean piracy that did not involve the use of physical media (such as CDs, DVDs, and flash drives) as the means of passing music, film, and other content from pirate to consumer. "Physical" piracy was covered in Phase I of the project (see OECD, 2008).

Structure of the study

This study is structured as follows. Chapter 1 presents markets for pirated digital products and discusses their economic properties. Chapter 2 explores factors that drive supply and demand of digital piracy. Chapter 3 presents industry actions against digital piracy and outlines institutional remedies to counter this phenomenon. The last chapter contains a case study of digital piracy affecting sports rights owners. This book has three annexes. Annex A outlines the legal foundations for copyright and their infringement. Annex B presents the 2004 OECD Council Recommendation on Broadband Development. Annex C contains the 2008 Seoul Declaration for the Future of the Internet Economy.

Definitions and parameters

The precise definition of "digital piracy" is fundamental to this analysis as this term can be interpreted in a number of different ways. In this study the term "digital piracy" means copyright infringements that do not involve the use of physical media (such as CDs, DVDs, and flash drives) as the means of passing music, film, and other content from pirate to consumer.

There are four points related to this definition of digital piracy that should be kept in mind.

- First, in usual computer terminology the term "digital" generally applies to the format in which data is stored, used and manipulated in computers, CDs, DVDs, etc. In this context it is the opposite to the term "analogue" which is used to describe data stored in, for example, vinyl records. Therefore, while the use in this study of the term "digital content" refers to data presented and stored digitally, the term "digital piracy" is used in a very specific way that is unrelated to this general common usage and refers specifically to infringements of intellectual property rights, that *do not* involve the use of physical "hard media" (such as CD, DVD, flash drives, etc) for the reproduction and exchange of pirated material. Therefore, in the context of this study "digital piracy" covers only Internet and direct computer to computer transfers, LAN file sharing, mobile phone piracy and so on.
- Second, in the context of this study the term "piracy" is used to describe infringements of copyrights and related rights[1] only. While the IP rights described in the WTO Agreement on Trade-Related Aspects of Intellectual Property Rights (TRIPS) – which form the basis of the OECD's work on counterfeiting and piracy, could

potentially cover the digital infringement of trademarks and industrial designs, this study focuses on digital infringements of copyrights and related rights, which parallels the way the term "piracy" is used in TRIPS.[2]

- Third, while the term "digital piracy" as defined earlier is used liberally in this study, there are some differences between economies as to what is considered as a copyright infringing activity and what can be considered to be fair dealing, fair use and private and domestic use. This is because copyright as a legal right is given effect through national laws. While these national laws have many similarities, either because they result from similar legislative interests that result in the establishment of copyright protection, or because they result from the international treaties on copyright that set minimum standards for that protection, as a general rule, countries are entitled to a certain level of flexibility in the implementation of their treaty obligations. Countries may provide a level of copyright protection that may exceed the treaty minimum. A more detailed review of legal foundations for copyrights, and acts constituting its infringement in different economies, is presented in Annex A.
- Last, to our knowledge legal jurisdictions do not offer a formal legal definition of "digital piracy", so this term does not necessarily have a clear and unambiguous meaning outside of the context of this study. In particular, while in this study "digital piracy" should be synonymous with the "digital infringement of copyright", it is arguable that outside of the context of this study there could be a distinction between the two concepts, at least from the legal perspective.

To reiterate, for the purposes of this study, "digital piracy" is taken to be, as defined earlier, the copyright infringing transmission in a form that does not utilise "hard media" of any copyrighted material including recorded music, motion pictures, software, books and journals as well broadcast and recorded performances covered by copyright.

Notes

1. In other words, the rights of phonogram producers and performers.
2. See TRIPS, Article 61.

Reference

OECD (2008), *The Economic Impact of Counterfeiting and Piracy*, OECD, Paris, *www.oecd.org/sti/counterfeiting*.

Chapter 1

Market Overview

This chapter outlines the markets where digitally pirated products are exchanged. To do so, it i) *explores the key properties of digital products,* ii) *identifies sectors that supply products prone to digital piracy and indicates possible delivery methods,* iii) *illustrates how the markets for pirated digital products operate, and* iv) *outlines the economic mechanisms driving those markets.*

Concerning the descriptive analysis of copyrighted products that have been subject to digital piracy, a set of properties has been identified that make these products significantly different from pirated products that have resulted in tangible/physical goods. These features include: the disembodied and non-physical nature of digital products, their interdependence on hardware, the characteristics that allow their digital delivery and the "immediate" potentially unlimited geographic scope of the market. These features have certain consequences for the further analysis of the phenomenon of digital piracy.

Numerous industries produce, supply and distribute copyrighted products that can be digitally pirated. Currently, the threat of this kind of piracy concerns such business sectors as music, movies, television (broadcast, satellite and cable, including transmission of sport events), radio, software (business and entertainment, including video games) and publishing (magazines, newspapers and books).[1]

Products supplied by these sectors can be delivered either via physical media (*e.g.* optical discs, paper books) or digitally (*e.g.* via computer networks, the Internet or as television signals). Therefore, digital piracy is not the only type of IP infringement affecting sectors that supply copyright-protected digital products. Many suppliers also face the threat of physical piracy (*e.g.* pirated CDs and DVDs or pirated software packages). These other types of infringements were addressed in the first part of the OECD's study on counterfeiting and piracy (OECD, 2008a).

Pirated products can be digitally distributed through a number of channels, including direct user-to-user contacts (between family members or colleagues) or more broadly through the Internet. There are numerous methods of distribution using the Internet, including peer-to-peer networks, one-click hosting services, "warez" sites, streaming sites, leech sites, or through other Internet-based solutions such as file transfer protocols or USENET.

Concerning the economic mechanisms present in markets for pirated digital products, the specific characteristics of digital piracy are such that classical economic analysis (markets composed of demanding and supplying parties operating on the basis of price/quantity) is not consistently appropriate in markets involving digital piracy. In these markets, even though some pirates are able to generate profits, (for example, by charging for access to pirated material or through sale of advertising around the audiences that are attracted by free material) some transactions occur without any monetary transfer from consumer to supplier, thus price and quantity demanded are not always the key features that link both demand and supply sides of the market for pirated digital products. Moreover, while in normal markets the distinction between suppliers and consumers is unambiguous, and it is easy to draw a clear line of division between them, in the case of digital piracy there are numerous instances (involving peer-to peer networks) where supply and demand enter the feedback loop, complicating the analysis.

Thus, in order to take into account this possibility while illustrating the market mechanisms the analysis relies on features other than the classical setting of price and quantity, and focuses more on related transaction costs and other, non monetary motives that drive suppliers of pirated digital products. This results in a distinction between markets where pirated digital products are supplied against a payment and markets where the exchange does not involve any significant monetary transfer.

Key differences between tangible and digital products

The key property of digital products is their disembodied character. This results in a set of characteristics that make digital products particularly different from tangible goods.

These differences result in a significant dissimilarity which in turn implies major differences between the ways in which piracy occurs in digital compared to physical goods. The key differences, which could affect the economic mechanisms of IPR infringement of each type of product, are outlined below.

Marginal cost of reproduction

The main consequence of the non-physical form of digital products is their virtually negligible marginal cost of reproduction and their ability to be digitally delivered (discussed below). It means that that once created, a digital good can in most cases be reproduced with relatively little cost and effort. Moreover, in most cases a copy of a digital product offers a level of quality comparable to that of the original. By comparison, in the case of tangible products the reproduction of a given good involves certain positive marginal costs of production and supply (*e.g.* materials used for production, costs of packaging, shipping, etc.), and the end result may be quite different in quality from the original product.

Digital delivery

The disembodied character of digital products permits their digital delivery, for example, via the Internet or local area networks. Such digital delivery implies a significant reduction in the cost of acquisition of a digital product (*e.g.* through reduction or elimination of searching, transportation and storage costs) which in turn significantly facilitates the distribution process. Thus, digital delivery offers a certain degree of flexibility of delivery, which is not available to physical products such as handbags or t-shirts.

Market scope

The extensive development of global computer networks have enabled digital products to be instantaneously delivered to any part of the world (see *e.g.* BSA-IDC, 2007). This implies that as well as the market for digital products having a global scope, many of the barriers that constrain the exchange of physical products (*e.g.* costly transportation, import tariffs, etc.) do not affect the exchange of digital products. This presents important opportunities for suppliers and consumers for both legitimate and illicit suppliers, as both can significantly reduce their costs of delivery and enhance their production and distribution processes, which in turn enable suppliers to offer consumers lower prices.[2]

Hardware dependence

Hardware is a necessary support for the existence of all digital products (*e.g.* storage on a hard drive, optical disc, server or other device), and the potential utility that digital products offer to an end user cannot be derived without suitable hardware. Consequently, there are strong linkages and complementarities between hardware and digital products, and in fact trends in consumption of digital products follow closely the newest technological developments (OECD, 2008b). While a similar circumstance exists for some physical products (for example a movie on a DVD will require hardware to play it), in most cases the counterfeited or pirated product exists in its own right (for example, a T-shirt, or a paper book).

Life span

In the current market, due largely to consumer tastes, content (especially digitised content) has been shown to have a shorter product life span than traditional physical goods have had in the past. Some research indicates that the demand for a particular digital product such as some music or movie shrinks much more quickly than demand for "classical" physical goods (Burke, 1996; Bhattacharjee *et al.*, 2007; Krider *et al.*, 2005).[3] The life span of some software products is also relatively short, as underscored in a report by the US Federal Trade Commission (FTC, 2003). Television shows have even shorter life spans, and in some cases can lose their attractiveness a few hours after transmission. The most extreme examples are live transmissions of sport events, in which their greatest attraction (and therefore value) occurs during the actual live transmission, and falls away very rapidly afterwards (see case study on the sports broadcasting sector in Chapter 4).

On the other hand the digitisation significantly increases the lifespan of digital products and extends their durability. Besides, some digital products have as long or even longer life span as classical tangible goods. For example, many classic video games from the 1980s and 1990s continue to have economic value today.[4]

Copyrighted digital products[5]

This section examines industry sectors that create copyrighted digital products and indicates possible delivery methods of those products.

Supplying industries

Modern economies tend to shift their focus from manufactured physical items to intangible assets. This shift results in an increase of importance of digital products[6] that are expected to provide new impetus for the economy, which is compounded by dynamic infrastructure developments such as the rapid growth of broadband (OECD, 2006). New technological developments, creative innovations and constant infrastructure improvements broaden and deepen existing markets as well as enhance potential access to a growing selection of digital products.

Today, digital products are key drivers of various high-tech industries including telecommunications, information technology, computers and home electronics. Simultaneously, digital products enter and change many existing sectors as new technological advancements emerge. Some digital products are offered as an alternative to the physical goods (*e.g.* music downloads for CD's; online movie streaming for a DVD movie). There are also completely new products that are uniquely digital and have no physical alternatives (*e.g.* online computer games, ringtones and entertainment and business software).

Available estimates in the recent *OECD Information Technology Outlook* (OECD, 2008b) indicate that industries that supply digital content report sizeable revenues across all sectors, with varying rates of growth and shares of overall revenues. For example, the global sales of interactive entertainment software amounted to USD 41.9 billion in 2007 (PWC, 2008). The online segment of this market is growing at the higher rate of 28% (OECD, 2008b). The film and video industries reported revenues of USD 84 billion in 2007, with a 5% growth rate, whereas the music industry had revenues of USD 29.9 billion and reported a negative growth rate of −10% (IFPI, 2008). The varying growth rates reported by the industry were the subject of several analyses, and the *OECD Information Technology Outlook 2008* lists digital piracy as one of the main threats to these industries that could significantly affect their growth rates.

Distribution channels

There are two ways digital products can be distributed – on physical media or digitally. The term "physical media" refers to a hardware good containing a given digital product. It can be an optical disc containing a movie, or a music album or a video game. As outlined earlier, digital delivery implies a non physical transmission of a given digital product, in particular through the Internet. A notable technological development that boosts Internet coverage is widespread broadband access (Figure 1.1).

Not unexpectedly, access to the Internet is rapidly growing in places where broadband as a communication and content distribution platform has evolved significantly.[7] Since 2003, broadband subscribers across the OECD have increased significantly (see Figure 1.2) with the most recent data indicating that by June 2008 these had reached 251 million (an increase of around 240%). Connections are faster and less expensive, with the average speed of advertised connections at almost 9 Mbit/s in 2007. Additionally, the introduction of, and emergence of, fibre networks additionally increases the total speed of transfer (see OECD, 2008c).

Figure 1.1. OECD broadband subscribers per 100 inhabitants, by technology, December 2008

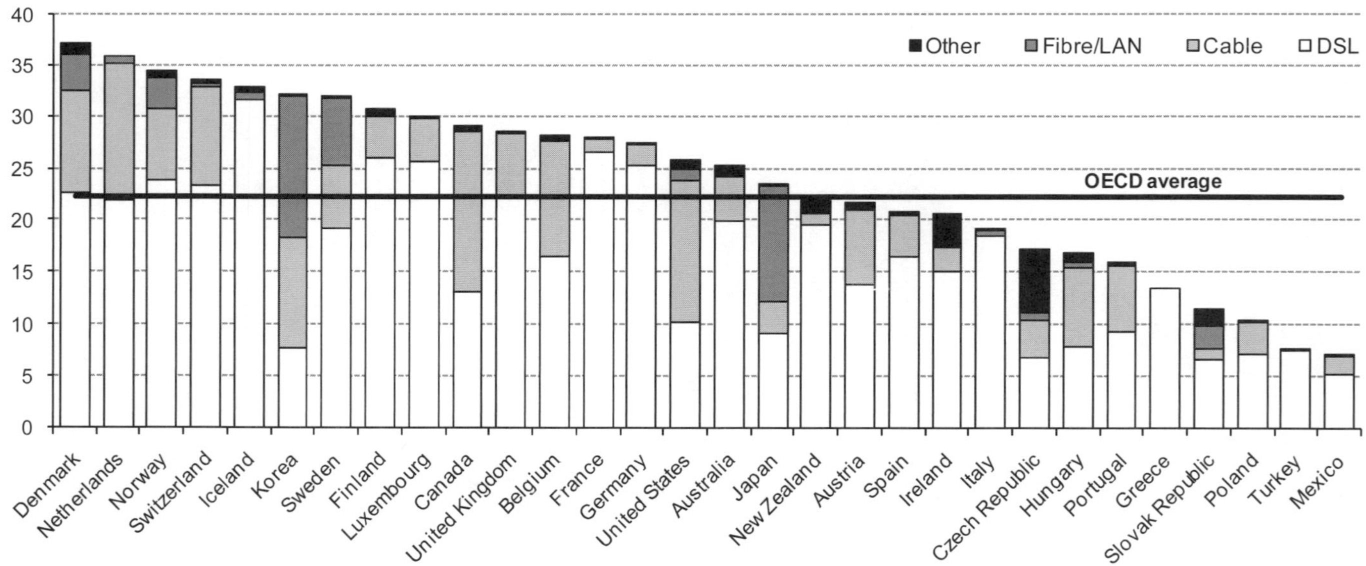

Source: OECD Broadband Portal, *www.oecd.org/sti/ict/broadband*, December 2008.

Figure 1.2. Broadband growth, OECD total, mid-2003 to mid-2008

Millions of subscribers

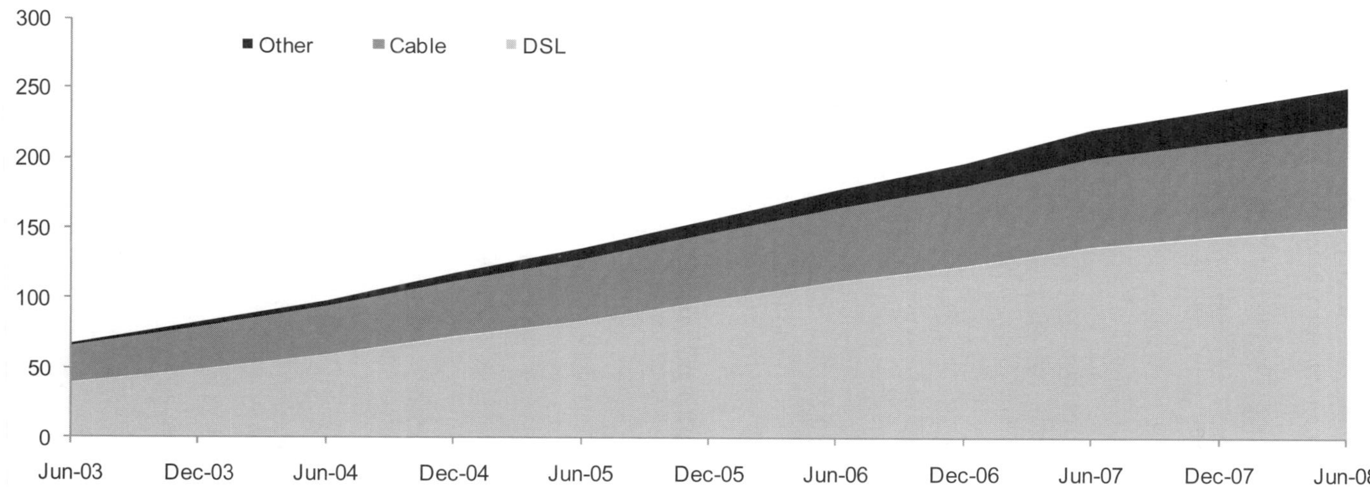

Source: OECD Broadband Portal, *www.oecd.org/sti/ict/broadband*, December 2008.

The importance of the emergence of the Internet goes well beyond the technical dimension and has numerous, profound socio-economic implications. Today the Internet is recognised as an essential platform for a knowledge-based economy and society. Technological and business innovation can be used to increase users' access to online content, while offering greater opportunities for numerous industries. This was also recognised by OECD member countries and is reflected in the OECD Recommendation on Broadband Development (see Annex B) and the June 2008 Seoul Ministerial Declaration on the Future Internet Economy.[8]

In the context of creative content industries, the Internet offers rights holders a choice over how their content is used and how it can generate new revenue opportunities. The Internet is not only providing more efficient distribution mechanisms, but is also generating new types of services and new forms of creativity that content creators can take advantage of. In fact the rapid growth of Internet coverage and saturation greatly promoted the emergence of new online business models employed by copyrights holders. These models, which rely on viable online delivery and the availability of large quantities of digital content, are rapidly gaining in popularity and their availability can be seen as an important part of confronting the threat of digital piracy (see Box 2.3 in the next chapter).

Markets for pirated digital products

This section presents the main technical ways in which markets for pirated digital products emerge. The exchange of pirated digital products takes place in several ways (presented below). Certain methods are web-based, whereas others do not require the use of the Internet. Some of these methods involve a monetary payment, whereas in others, consumers can access pirated digital content for free.

The hardware dependence of digital products implies that the growth of the personal computer market is closely linked with the markets for pirated digital products. As outlined in a study by Solutions (2006), a personal computer no longer acts as a working tool but has become "a life hub and video entertainment centre". This indicates that the widespread use of personal computers not only grants unprecedented individual access to any form of information, but also acts as a platform for the market for digital products (genuine as well as pirated). In fact the strong market for personal computers was indicated in the recent BSA-IDC report as the key for the development of markets for pirated digital products (BSA-IDC, 2007).

In addition to the dramatic increase in the percentage of households that own personal computers, the rapid development of the Internet implies that more people than ever before have access to practically any type of news or data. However, this technological progress also facilitates digital piracy, as users employ various web-based solutions and applications to distribute and exchange pirated digital products. Hence, a significant volume of digital piracy occurs via the Internet, which is the main way of exchange of all types of digital information, and which supports and facilitates the distribution of all types of digital content.

There are numerous ways pirated digital products are exchanged. Some of them are acquired from local sources (such as family members and colleagues through direct user–to–user contacts); others are distributed online, using the peer-to-peer (P2P) networks, or dedicated sites, such as "warez sites", one-click hosting services, leech sites, linking sites, cyber-lockers, streaming websites, etc., many of which appear to co-operate closely with P2P websites. The last possible method of digital piracy relies on other technical solutions such as file transfer protocol (FTP) or USENET.

Local sources

The easiest and often the fastest method for users to acquire pirated copyrighted material is using classical methods of exchange (described below) from local, direct sources such as friends, family members, colleagues from school, studies or work, etc.[9] As indicated by the latest British Music Rights survey on music consumption habits, consumers perceive this method of exchange of pirated digital products as the most natural and habitual (BMR, 2008).

Technically, there are several ways of distributing pirated digital products in a local, social network. The easiest way is to send a pirated copy to another user via local computer networks. However, even if two computers are not directly connected, the Internet offers a number of possibilities for direct exchange of pirated digital products. The increasing capacity of e-mail accounts and the growing size of permitted attachments mean that the direct sending of large files via e-mail is becoming increasingly easier, especially as many free e-mail services offer more than one gigabyte of free mailbox space. Software programmes used for "chatting" also support file transfer mechanisms between users. For larger files (*e.g.* music albums or movies) users turn to dedicated file uploading and storage services that offer large storage capacities, and web access from multiple locations. After the file is uploaded, a user receives a retrieval URL that can be further distributed to any intended recipient.

Copyrighted material can also be transferred without the use of a computer by using mobile phones or other devices, such as personal digital assistants (PDAs) or mp3 players equipped with network connecting components. In all these of these methods the user is acquiring directly pirated material from known, local sources, on a personal basis via the Internet or local network.

Peer-to-peer networks

Formally, peer-to-peer (P2P) technologies are defined as a "communication structure in which individuals interact directly, without necessarily going through a centralised system or hierarchy. Users can share information, make files available, contribute to shared projects or transfer files" (OECD, 2005a). Technically speaking, the term peer-to-peer denotes organised networks that connect computers across the Internet with special file sharing protocols implemented by a given programme. Computers are linked together in a P2P network with the purpose of interaction and file exchange. The primary function of P2P file sharing is a legitimate means of transferring information from one spot to another in an efficient and cost effective manner, and P2P is perceived as fundamental for further technological development that could lead to new, innovative solutions for information transfer.[10]

A particular feature of P2P networks that has become one of the reasons for their popularity is user participation. With a sufficient number of contributors, a P2P user can easily find and quickly download most of the information and digital products he or she demands, as the combined processing power and increasing transmission capabilities of networked computers provide unprecedented access possibilities to digital resources. However, as a downside, the peer-to-peer networks are also used as a way to infringe copyright and to distribute malicious software.

The P2P networks vary with respect to the degree of centralisation (OECD, 2005b). Centralised peer-to-peer networks depend on dedicated servers to record user information and accumulate pieces of the shared data. On the other hand, decentralised networks rely on individual users to provide information related to other users and the files that are being distributed. In most decentralised networks each computer is simultaneously sending and receiving information and each user is simultaneously a supplier and consumer of digital products.[11] This piece by piece sharing technique, through "encouraged reciprocity", has resulted in a significant increase of transmission speed.

Although the technology behind P2P file sharing was developed for legitimate purposes of quick and efficient information sharing and data exchange, networks that use the technology have emerged and now they are used to distribute virtually all types of digital content, including pirated music, movies and software. In addition, recent technological developments also permit P2P networks to also be used for the unauthorised distribution of live broadcasts of sports events, which has become a significant threat for the sports broadcasting industry (see Chapter 4). Examples of peer-to-peer networks and protocols include GNUtella, eDonkey and BitTorrent.

The popularity of the P2P networks is reflected in available data. Since users are increasingly downloading larger files (*e.g.* movies, software or bundled music albums), the peer-to-peer networks generate a significant volume of Internet traffic, that according to Ellacoya (2007) recently amounted to 39% of the total Internet traffic in North America (Figure 1.3). Other studies suggest this number was much higher; according to the consulting company Ipoque between 49% to 83% of Internet traffic was generated by P2P file distribution in 2007 (Ipoque, 2007).

Figure 1.3. Internet traffic in North America by category, 2007

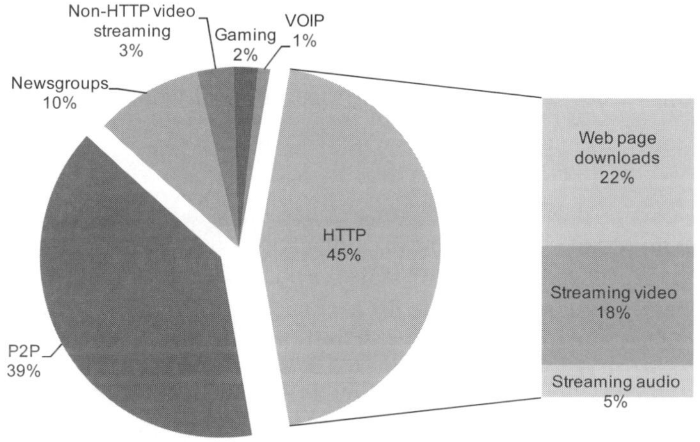

Source: Ellacoya (2007), "Ellacoya Data Shows Web Traffic Overtakes Peer-to-Peer (P2P)", media alert, 18 June, *www.ellacoya.com/news*, accessed 28 February 2008.

Some P2P networks were often also reported to be a channel of distribution of various types of potentially harmful malicious software such as trojans, viruses, etc. For example, the security test performed by the service "Site Advisor" indicated that digital content acquired through one of the most popular P2P networks could expose a downloading consumer to a threat of spyware or other potentially unwanted programs.[12]

"Warez sites" [13]

A broad spectrum of pirated computer programmes, music, movies, books and other copyrighted content is published on the so-called "warez sites" in an organised but illegal fashion. Organised groups, often called "release groups", specialise in delivering new, often unreleased music albums, movies or other media on selected websites – or so-called FTP sites – where users can access large quantities of illegally copied content. These sites are sometimes used as servers to stream pirated content such as television shows or live sports events. Besides distributing pirated movies, music and other content, "release groups" are also known for the illegal distribution of licensed software. These services are often offered against a payment by end users; but the price charged by the suppliers of pirated content is mostly much smaller than the price of the original products. Similarly to misused P2P networks, the "warez sites" are also reported to be a source of malicious software such as trojans, viruses, etc.

"One-click" hosting services

Another method of distribution of pirated digital content is via the so-called "one-click" hosting sites also known as "file" or "cyberlockers" that allow Internet users to upload easily and free of charge one or more files from their hard drives onto the one-click host's server. Most of these services simply return a link to an Internet page (URL) which can be transferred to other people, who can later download the file. Many Internet forums exist in order to share such links and in addition to legal content a vast number of files that are hosted on these services contain pirated digital products such as movies, music, software or recorded television shows, and sport transmissions.

Online sharing and streaming websites

Online media, video, music sharing and streaming websites, including user-created content sites, have become a popular method of distribution of digital content, in particular audio and video content. According to OECD (2008b) the user-created content sites are a big source of sharing of digital content, including, for some, unauthorised sharing. Today these services are some of the most frequently accessed online services among Internet users, and are often used by rights holders to distribute their content (OECD, 2007). According to the Ellacoya survey results, the HTTP traffic recently exceeded P2P, mostly due to "a surge in the use of streaming media, mostly video" (Ellacoya, 2007).

In the context of digital piracy, there are two ways intellectual property infringement may occur via online sharing and streaming: *i*) websites that offer direct access to the copyrighted material, or *ii*) through re-use of an IP-protected product in a user created content beyond fair and personal use permitted by law.

Direct sharing refers to websites that offer online access to content such as video, music or rebroadcasted television programmes (such as transmission of sports events). In this way the whole, or part, of a copyrighted digital product, such as a movie, music song or a broadcast TV programme, can be uploaded to an online sharing service for public access. This method also includes the capturing and unauthorised re-transmission of broadcast signals through so-called Unicast technology (transmission to a single user, as opposed to "broadcast" – transmission to multiple users), which occurs mostly in the context of pirated retransmission of live sports events.

The unauthorised use of copyrighted digital content, beyond fair or other (*i.e.* personal) use permitted by law,[14] sometimes occurs in the context of the creation and dissemination of user-created content.

Leech sites

Leech sites[15] have quickly grown to be a primary concern for content providers worldwide. This technique works in two steps: it starts with the illegal uploading of content on a site which is designed for media generated by end users (user-generated content, or UGC site)[16] usually with a name that will not attract attention. Then, on a leech site, a display is made of the true description of the content that was uploaded in step one, along with a link to the direct, exact address of that content on the UGC site. When selected on the leech site, the content thus starts playing immediately in a window belonging to the leech site environment, therefore masking the UGC site where the content is concealed.

While leeching is technically preventable, source sites (video hosting sites including UGC sites), are less inclined to take action as high volumes of users add to the site's desirability. In addition, leeching causes the source site's web server to respond to a lot of leech requests which will use up the server's bandwidth, potentially resulting in:

- The source site's web server slowing down, causing performance to lag.
- The source site's web server's bandwidth consumption increasing. For smaller sites, this can be an expensive problem if it reaches their hosting plan's bandwidth limit and they have to subscribe to a more expensive hosting plan.

- The source site can lose ad revenue it would have gained if web surfers directly access their site.

A somewhat related method is the so-called deep linking or making a hyperlink, which refers to the practice of inserting a link pointing to a specific page on another website. Links or hyperlinks are central for web functioning usability and navigation. However, the practice of websites dedicated to providing links to pirated content can have a destructive effect on rights holders.

Other means of distribution

Other methods of distribution of pirated digital products include more conventional means, such as file transfer protocol (FTP), web servers, Internet Reality Chat (IRC) or Usenet.

The administrators of web servers and FTP servers have the ability to publish copyrighted content accessible by all Internet users. However, these methods of distribution are much slower than the dedicated file hosts described above. Moreover, (unlike the P2P networks) a sudden increase in web traffic during a short period of time can easily cause a system breakdown, which makes this method impractical and thus unpopular for a large set of users. However, in some situations, operators of FTP servers use the infrastructure of third parties, such as universities, hospitals, public libraries or even homes for the elderly, in many cases unbeknownst to them. In these cases high data transfer speeds can be obtained as many of these organisations now have fibre-optics networks available.

Certain copyrighted material could be also exchanged using IRC Channels and Usenet. The latter, more particularly, has become very popular in recent years, as Internet users no longer need specialised knowledge, and with only the Usenet infrastructure being used for file sharing distribution, rather than the newsgroup environment. This has become evident, with Usenet providers now offering HTML-based search engines that allow subscribers to search the Usenet through their web browsers, list all available files ranked by their reliability and popularity and then download them with exceptionally high speeds.

Economic mechanisms that drive markets for pirated digital products

This section illustrates market actors active in markets for pirated digital products and outlines the main economic mechanisms of these markets.

Market actors

Existing analytical studies suggest that in most societies the distribution of actors active on the markets for pirated digital products is highly skewed towards young males.

A study by Gopal and Sanders (1998) that analysed behaviour related to software piracy, found that gender and age are correlated with this behaviour, and that young males form the group that is the most likely to engage in software piracy. These results were confirmed by Bryce and Rutter (2005), who portrayed an average person that exchanges pirated digital products as being male and below the age of 30. The finding about the youth of a "representative pirate" was also presented in the study by Solutions (2006) which found that the most intense IP infringements in the United States occurs in the group of people aged between 12 and 20 years (18% of the US online population).

Lenhart *et al.* (2007) provided a partial explanation to the question of why young people are more likely to digitally violate the IP rights, by reporting that the use of media is central to many teenagers' lives. This observation was confirmed by another study by Wiggin (2008), who found that individuals that engage in digital piracy are typically heavy media consumers. Given that young people generally have relatively low incomes, digital piracy becomes for them a possible way of getting what they want without paying for it.

The 2007 study by Lenhart *et al.* also provided more insight into the demographic structures with respect to different types of markets for pirated digital products. The analysis, performed in the United States, suggests that young women and men are almost equally likely to use video sharing websites and streaming services, whereas men are more likely to use other services, and are generally more likely than women to upload pirated digital content.

Economic mechanisms

In traditional markets where goods are exchanged against a monetary payment, it is appropriate to start the analysis of market mechanisms by using the classical supply and demand perspective. According to this perspective, the demand and supply sides are linked together with the monetary payment, at a given price, for the provision of the good or service. The supplier is the party that offers the good or a service to the demanding party that pays a given price for it. The production or delivery of the good or service involves certain marginal costs, which implies that the supplier demands a monetary payment in exchange for the delivery of the good or service. If the price is too low (in particular if it is lower than the marginal cost of production and delivery) then the supplier experiences monetary losses.

However, in the case of digital piracy a different situation is observed. In particular there are two phenomena that cannot be explained by a direct application of the traditional setting.

- First, some economic agents supply pirated digital products at zero price.
- Second, some consumers are willing to pay certain prices for legitimate digital products, even if they can possibly choose a pirated copy at no price.

In fact, the structure of markets for digital products is quite unique from an economic perspective, since there are three co-existing and competing markets where digital gods are exchanged: the legal market that is carried out on the principle of monetary payment and two illegal submarkets: in one of them the pirated digital products are distributed at no price in the other the exchange is made against a monetary payment (Box 1.1).

The instances described in Box 1.1 represent markets which have been facilitated by technology. In order to describe the mechanisms of such markets, an extension of the simple, traditional approach is required, where the rationale for supplying on the market may be different from the prospect of a profit stream, and demand is driven by factors other than the price consumers need to pay. From an analytical point of view, two other groups of factors need to be introduced to the analysis in order to understand the mechanisms that drive markets for digital products:

- On the supply side there are non-monetary drivers (*i.e.* factors other than prospect of profits) that motivate some market actors that supply pirated digital products at no price.
- On the demand side there are transaction costs apart from price that drive consumers' decisions. These transaction costs include all types of socio-economic costs other than price that are incurred by a consumer in the process of acquisition of a given digital product.

Box 1.1. Economic markets for digital products

On markets for legitimate digital products, the suppliers are the legitimate copyrights holders and the digital products are exchanged against a given price. The production of a digital good involves certain costs, which implies that the supplier demands a monetary payment in exchange of delivery of the good. If the price is too low to cover the costs of production of the good the supplier experiences monetary losses.

Concerning pirated digital products, two types of markets can be distinguished: the non-monetary submarket and the monetary submarket.

- In the non-monetary submarket, the exchange of pirated digital products occurs free of monetary charges and supply is often driven by factors other than the prospect of monetary profits (although some economic agents that supply pirated digital products at zero price can earn income from advertising). This way of exchange can be observed in the cases of: one-click downloading services, P2P networks, leech sites, online sharing websites and some streaming websites.
- In the monetary submarket, a user pays a certain monetary amount in order to gain access to a pirated digital product, for example some products on warez sites (mostly software) that are offered against a price which is usually much lower than the price of the legitimate product. Certain on-line streaming services also require a monetary payment in order to grant the access for a consumer to the pirated digital content. In this case a payment results in access to a streaming channel that provides a given pirated digital good (frequently sport events). The scope of the monetary submarket seems to be rather limited since it is in competition with the market for legitimate products and pirated products offered at no price. Nevertheless, it is highly probable that suppliers of the pirated digital products enjoy exceptionally high profit margins from their activity, since duplication of the digital product would be at little or no monetary marginal cost.

Non-monetary drivers

Free distribution seems to dominate the markets for pirated digital products, although advertising driven models for legitimate products are also growing (but hampered by the availability of pirated products.) All of the previously cited technical ways that facilitate market formation include exchange methods that do not involve monetary payment. This applies to local sources, P2P networks, non-participating web-based services, leech sites and other methods of exchange.

In order to understand how a systematic delivery of a product free of monetary charge is possible it is important to recall that the marginal cost of replication of digital products is in most cases virtually zero. In fact the reproduction of most digital products is easy, requires little time and effort and does not involve the use of any complicated equipment. In all these cases, the total non-monetary marginal cost of supply of pirated digital products equals the cost of time devoted to duplicate, plus a given psychological cost of unlawful behaviour and expected penalty (taking into account the risk of discovery).

Given that the marginal cost of supply of pirated digital product is small, a potential infringer does not require a significant incentive in order to begin supplying, and can earn profits from sources other than customers' payments (*e.g.* advertisements). Other motives that drive supply of "free" pirated digital content include permissive attitudes towards digital piracy that often result in recognition of a supplier within a social group, or a "reciprocity mechanism". These supply drivers that shape the market potential are presented in greater detail in the following chapter.

Transaction costs

The co-existence of markets for legitimate digital products (where products are exchanged against a monetary payment) and markets for pirated digital products where products can be acquired at no price) indicates that the demand for these goods can be largely influenced by factors other than price. In other words, since some economic agents decide to buy digital products and pay an amount of money for them, even though a free, pirated alternative can be found on the market, it must imply that there are factors other than price that drive their decisions. These factors can be classified as transaction costs related to acquisition of a given digital product.

In economic terms, a transaction cost is something that is paid in order to enable an economic exchange. The term "transaction costs" was first introduced by Coase (1937) who analysed their importance for the efficiency of a firm's operation. The transaction costs were then successfully applied in numerous ways in order to study and explain a number of different market behaviours. There are numerous types of costs identified and described in the literature[17] including: costs of searching (time and effort spent in order to determine if a demanded good is available on a given market, what the price is and which supplier asks the lowest price, etc.), bargaining costs (cost of coming to an acceptable agreement with the other party), policy costs (costs of supervision if the other party fulfils the agreed terms of the contract) and costs of enforcement (cost of legal action if the other party does not fulfil the contract).

In the context of the digital products market the transaction costs can be applied in order to evaluate factors that can affect consumers' decision whether or not to acquire a given digital product. Remarkably, the transaction costs apply to all types of products on the market (not only to the IP-infringing products). All transaction costs that are related to acquisition of a legitimate product act as drivers that potentially promote digital piracy, and all transaction costs related to acquisition of a pirated digital product act as negative drivers of digital piracy that reduce its scale. The transaction costs related to legitimate and pirated digital goods are briefly summarised below. A more thorough discussion is presented in the next chapter.

Concerning legitimate digital products, the most prominent transaction costs include: *i*) searching costs, *i.e.* all factors that affect the ease of finding and acquiring a given legitimate product that refer to the time and effort spent on searching; and *ii*) the cost of the overall package, which refers to the case when a given digital product is offered in a bundle. This restrains a consumer from picking a single object and forces him or her to pay more money for those products he or she does not want. All these costs will therefore reduce the attractiveness of a legitimate product and will therefore indirectly act as demand drivers for digital piracy. See Chapter 2 for a more detailed discussion.

Regarding the markets for pirated digital products, there are various types of transaction costs that a potential consumer expects to incur. Just like on the markets for legitimate products, there are searching costs that determine the perceived availability of the pirated digital products (*e.g.* time devoted to searching, degree of technological literacy required to download a given product, etc.). Product-related transaction costs include possible safety hazards involved in downloading pirated digital content (*e.g.* viruses, malware) and cost of potential customisation of a pirated copy (*e.g.* of a pirated software product that requires tailoring). Moral transaction costs capture the ethical disutility caused cost of engaging in an illicit activity. Finally, legal transaction costs include the fear of possible prosecution and penalty, etc.

Markets for pirated digital products

The importance of transaction costs is the key element that in turn leads to the emergence of two submarkets, where the pirated digital products are exchanged: monetary and non-monetary submarkets (see Box 1.1). This subsection investigates the necessary conditions for these markets to emerge.

In order to understand the mechanisms that lead to the emergence of these markets, it is useful to look at the microeconomic conditions that affect a consumer's decision as to whether to acquire a given product (including digital content). Generally, there are three general conditions that need to be simultaneously fulfilled for a consumer to demand a given product:

- First, a consumer must expect to derive a certain utility from consumption of this product. It means that a consumer must like the product in a way that the satisfaction derived from consumption offsets the price demanded (if any) and the involved transaction costs.
- Second, a product must be within the consumer's budget constraint, *i.e.* the product must be affordable and a consumer must have enough disposable income in order to pay for it.
- Third, a given product must be preferred over alternative ones. In other words the product that a consumer chooses is better than any other alternative on the market in terms of price and transaction costs.

Legitimate vs. pirated products

A consumer that demands a given digital product can get a legal copy (that may or may not involve paying a price), or decide to search for an illegal pirated version. He or she will prefer a legal copy if the total perceived cost of purchase (*i.e.* price plus transaction cost) of a legitimate copy is lower than the transaction cost of acquisition of a pirated copy. This could happen if the transaction costs for a pirated copy are particularly high and if the total costs of acquisition of a given legitimate product (price plus transaction costs) are affordable for the consumer.

The transaction costs for a pirated copy may be particularly high when for example a consumer is technically illiterate, the quality of the pirated product is suspect or expected penalties are high. A consumer is also more likely to opt for a given legitimate product if its price makes it affordable, and the product can be easily accessed. This reinforces the case made in previous related OECD studies on digital content (OECD, 2005a, 2005b), which suggest that the development of innovative business models for legal digital products (including advertising-based models or models based on new delivery methods over the Internet) may affect the economic rationale behind the attractiveness of digital piracy. However, that being said, legitimate products, which also incur significant costs, including licensing fees, digitisation and configuration costs, taxes, etc, will never be able to compete on a level playing field with illegal products offered free of charge that do not incur such costs. Therefore, innovative business models may only address a fraction of the overall problem.

Emergence of markets for paid pirated products

A comparatively high price for a legitimate product or its unavailability on a particular market, as well as the availability of illegal copies can trigger certain consumers to seek pirated alternatives. A consumer that decides to acquire a pirated digital product will prefer to do it at no price, rather than paying anything for it, unless the transaction costs are judged to be too high in which case the consumer may consider a paid illegal product in order to overcome those (non-monetary) transactions costs. This could occur, for example when a given digital product is rarely sought by consumers and cannot be easily found on the free illegal market (*e.g.* a "niche" music album or movie that is shown in theatres only); or its duplication and customisation requires a high degree of technological literacy (*e.g.* specialised software products); or the delivery of the product is technologically difficult (*e.g.* live streaming of sport events). Ultimately, if the transactions costs associated with free goods are excessive, and the price of a pirated product on the illegal monetary market is too high compared to the price of a legitimate good, then a consumer might choose to buy the legitimate good or not to buy at all.

To reiterate, there are two conditions that favour the emergence of a monetary submarket where pirated products are exchanged against a monetary payment: *i*) there is no known "free" pirated alternative or the transaction costs related to its acquisition are too high, and *ii*) the legitimate product is judged to be relatively too expensive or is unavailable for the individual consumer.

Figure 1.4. Markets for pirated digital products

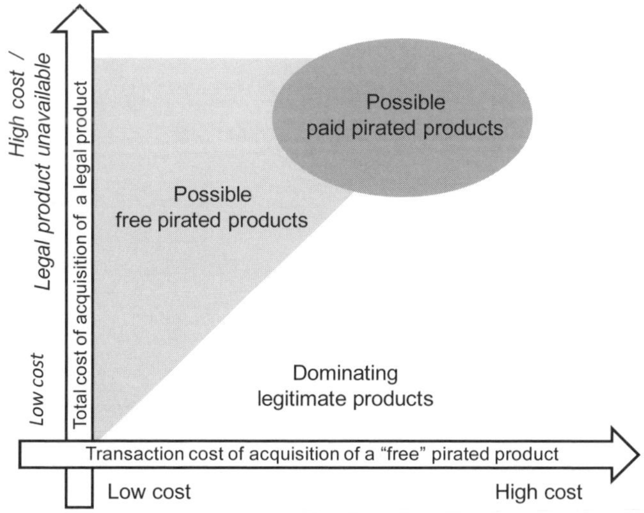

Figure 1.4 illustrates the conditions under which the markets for pirated digital products are likely to emerge. The vertical axis represents the total cost of acquisition of a legitimate product (its price plus related transaction costs). The horizontal axis represents the expected transaction costs of acquisition of a free pirated product.

A consumer may prefer to pick a legitimate product, as long as its total cost of acquisition (price and transaction costs) is lower than the expected cost of acquisition of a pirated alternative (which corresponds to the bright area on the figure). Once the total costs of acquisition of a given legitimate product become high, the pirated market is likely to emerge. The "free" pirated market will dominate as long as the transaction costs related to the acquisition of a free pirated copy remain relatively low (which corresponds to the light grey area on the figure). If the expected transaction costs of acquisition of a free copy are high and the legal product is expensive or unavailable, the "paid" illegal market is expected to appear (which is corresponds to the dark grey circle in Figure 1.4).

Notes

1. For OECD sector studies on digital content, see *www.oecd.org/sti/digitalcontent.*
2. Clearly this affects only the distribution costs and not fixed costs of creation of a given digital products paid by rights holders.
3. Bartlett and Snelus (1980) identified that for each generation of consumers only a few selected music titles can have a long-term life span.
4. For example, in 2007, the top-selling game on Xbox LIVE Arcade, the online marketplace for games for the Xbox 360 game console, was Teenage Mutant Ninja Turtles, a 1989 arcade game. See Grant, Christopher, Top Xbox Live, Arcade titles of 2007, Joystiq (4 January 2008), *www.joystiq.com/2008/01/04/top-xbox-live-xbox-live-arcade-titles-of-2007.*
5. See also *www.oecd.org/sti/digitalcontent.*
6. The term "digital product" is often used in reference to a broad set of intangible tradable goods and services. Many digital products potentially could be subject to digital piracy.
7. The OECD Council issued a recommendation on broadband development in 2004. See Annex B.
8. See OECD (2008d).
9. Some of these cases can be classified as the so-called personal or fair use which to a varying extent is permitted by law in many economies (see Annex A).
10. For discussion on the future of P2P solutions see: *skype.com/help/guides/voip/*, *www.p2p-next.org* and *distribution.openoffice.org/p2p/index.html.*
11. In some P2P networks a user can disable the uploading function so that the system only allows incoming traffic.
12. Report on Kazaa on *www.siteadvisor.com*, accessed 28 April 2008.
13. Initially the term "warez" referred to pirated software distributed over the Internet. It now tends to denote a broad spectrum of online pirated content including music, movies, software, etc.
14. See Annex A and OECD (2007).
15. Also known as hotlinking, direct linking, bandwidth theft, video linking or video leeching.
16. User-generated content (UGC) is also known as user-created content (UCC); see OECD (2007).
17. See Williamson (1981) for an overview.

References

Bartlett, J.C. and P. Snelus (1980), "Lifespan Memory for Popular Songs, *American Journal of Psychology*, 93(3).

Bhattacharjee, S., R.D. Gopal, K. Lertwacharab and J.R. Marsdena (2007), "Stochastic Dynamics of Music Album Life Cycle: An Analysis of the New Market Landscape, *International Journal of Human-Computer Studies*, 65(1).

BMR (2008), Music Experience and Behaviour in Young People, British Music Rights Organisation, available at *www.ukmusic.org*.

Bryce, J. and Rutter, J. (2005), "Fake Nation?", available at *www.allianceagainstiptheft.co.uk.*

BSA - IDC (2007), "Fourth Annual BSA and IDC Global Software Piracy Study", Business Software Alliance, available at *www.bsa.org/globalstudy,* accessed 6 March 2008.

Burke, A.E. (1996). "The Dynamics of Product Differentiation in the British Record Industry", *Journal of Cultural Economics*, 20(2).

Coase, R.H. (1937), "The Nature of the Firm", *Economica* 4.

Ellacoya (2007), "Ellacoya Data Shows Web Traffic Overtakes Peer-to-Peer (P2P)", media alert, 18 June, *www.ellacoya.com/news*, accessed 28 February 2008.

FTC (2003), "To Promote Innovation: The Proper Balance of Competition and Patent Law and Policy", a report by the Federal Trade Commission, USG.

Gopal, R.D. and G.L. Sanders (1998), "International Software Piracy: An Analysis of Key Issues and Impacts, *Information Systems Research*, 9(4), 380-397.

IFPI (2008), "IFPI Recording Industry in Numbers", International Federation of the Phonographic Industry.

Ipoque (2007), "ipoque Internet Report 2007: The Impact of P2P File Sharing, Voice over IP, Skype, Joost, Instant Messaging, One-Click Hosting and Media Streaming such as YouTube", available at *www.ipoque.com.*

Krider, R., Tieshan Li, Yong Liu and Charles B. Weinberg (2005), "The Lead-Lag Puzzle of Demand and Distribution: A Graphical Method Applied to Movies", *Marketing Science*, 24(4).

Lenhart, A.M, M., A. Rankin Macgill and A. Smith, (2007), "Teens and Social Media: The Use of Social Media Gains a Greater Foothold in Teen Life as Email Continues to Lose its Luster", Washington, DC: Pew Internet & American Life Project, 19 December.

OECD (2005a), "Digital Broadband Content: The Online Computer and Video Game Industry, DSTI/ICCP/IE(2004)12/FINAL, available at *www.oecd.org/sti/digitalcontent*.

OECD (2005b), "Digital Broadband Content: Music", DSTI/ICCP/IE(2004)13/FINAL, available at *www.oecd.org/sti/digitalcontent,* OECD, Paris.

OECD (2006), *OECD Information Technology Outlook 2006*, OECD, Paris.

OECD (2007), *Participative Web and User-Created Content: Web 2.0, Wikis and Social Networking,* OECD, Paris.

OECD (2008a), *The Economic Impact of Counterfeiting and Piracy*, OECD, Paris.

OECD (2008b), *OECD Information Technology Outlook 2008*, Chapter 6, "Digital Content", OECD, Paris, *www.oecd.org/sti/ito.*

OECD (2008c), *Broadband Growth and Policies in OECD Countries*, OECD, Paris, available at *www.oecd.org/sti/ict/broadband/growth*.

OECD (2008d), "Shaping Policies for the Future of the Internet Economy", OECD Ministerial Meeting on the Future of the Internet Economy, available at: *www.oecd.org/futureinternet.*

PWC (2008), *Global Entertainment and Media Outlook: 2008-2012*, Price Waterhouse Coopers.

Solutions (2006), "Movie File-Sharing Booming", Solutions Research Group, available at *www.srgnet.com*, accessed 28 March 2008.

Williamson, O.E. (1981), "The Economics of Organization: The Transaction Cost Approach", *The American Journal of Sociology*, 87(3), 548-577.

Chapter 2

Drivers of Digital Piracy

This chapter examines the supply and demand drivers of digital piracy.

Supply drivers

The decision of a party to supply pirated digital products is driven by factors related to *i*) the market potential that include the personal motives that drive suppliers of the pirated digital content, *ii*) technological considerations that facilitate reproduction and distribution of pirated digital content, and *iii*) the institutional environment that determine relevant market risks (Table 2.1). These factors may differ according to the type of infringement involved, as well as across different economies.

Table 2.1. Supply drivers of digital piracy

Driving factors	Characteristics that favour digital piracy
Market potential	
Attitude towards piracy	Permissive attitude towards digital piracy, social recognition, collectivism
Reciprocity mechanism	Mechanism imposing reciprocal behaviour
Profitability	Relatively high potential profitability
Market scope	Large potential market
Market exploitation (technology)	
Technological developments affecting:	
• reproduction	Easy and costless reproduction
• distribution	Rapid growth of the Internet, simple and quick distribution
Market risks	
Legal and regulatory framework	Permissive law, inefficient enforcement
Parental control	Lack of parental control

Market potential

In contrast to normal economic activities, parties that engage in the supply of pirated digital products do not always do so to make a profit. In many cases, the supply of pirated content depends on other potential types of rewards determined by the attitude towards piracy present on a given market or the presence of a reciprocity mechanism. If monetary payment is involved, the attractiveness of digital piracy depends mostly on the expected profitability.

Attitude towards digital piracy

An individual's attitudes towards digital piracy could influence the extent to which he or she supplies pirated digital products. In particular, the mechanisms in some types of market for digital pirated content could lead to the perception that supplying digital piracy might not be an illicit or blameworthy activity, especially as a significant part of the exchanges of pirated digital products occur without profit motives, which can be perceived as socially acceptable "sharing".

Social recognition seems to be particularly important in the case of pirated digital products distributed via web services located in the http space (such as warez sites, online video and music sharing and streaming websites, or some user-created content sites). In all these cases, pirated digital content is freely available and can be accessed at no cost by any visitor on a given site or service. By introducing pirated digital products to the market an infringing party expects to get recognition in certain groups, either by breaking the copy protection (in case of software offered on warez sites), by using a given copyright protected product in creating other material, or simply by marking a given digital product with a personalised label or symbol.

Surprisingly, this permissive attitude seems to change, once the problem of digital piracy affects user-created content. According to the study by Lenhart *et al.* (2007) most young users of the Internet (*i.e.* the most active group on the market for pirated digital products) restrict access to their posted content, at least some of the time.

From a macro perspective, collectivism[1] was a part of the social environment identified to be strongly and positively correlated with digital piracy rates. This result was established by a number of authors, who used Hofstede's cultural model (Hofstede, 1980, 2004) in order to explain the cross economy variation in digital piracy levels.[2] Hofstede's model aggregates various economy-specific cultural factors and is often used in studies on effects of economy-specific culture on various phenomena. Apart from collectivism, several studies pointed at the weaker effect of several other cultural factors on digital piracy, including masculinity, uncertainty avoidance and power distance.[3]

Reciprocity mechanism

The "reciprocity mechanism" implies that a user supplies a given pirated product because he or she expects to receive another product in the future (Fehr *et al.*, 1997; Fehr and Gächter, 2000). Thus, a single user can dynamically switch roles between supplier and consumer, and in some cases (*e.g.* by participating in some P2P networks), can act simultaneously as supplier and consumer.

The reciprocity mechanism drives supply in local sources, and to some extent in P2P networks and one-click hosting services. Since the term "local sources" refers to users' family, friends and colleagues, the exchange of pirated digital content in these networks occurs out of general self-gratifying, reciprocity motives, where a given person offers a pirated product in order to maintain a social relationship.

Concerning pirated content exchanged via P2P, the technical construction of these networks permits and encourages users to supply and demand simultaneously. Thus these exchange methods promote "reciprocity" from participating users. Suppliers of pirated digital content on one-click hosting services usually expect to gain better access to other users' pirated digital products. This improved access can be gained in two ways.

Most of the one-click hosting services offer better access to their service to those users that upload popular content. Thus, an upload of a popular digital product (*e.g.* a movie or music album) increases the users' chances of receiving a technically improved access to the service in the future.

Structured and easily accessible information on pirated digital content available via one-click hosting services is available on numerous Internet forums, which are often restricted to registered and active uploaders. Hence, access to such information can be perceived as reward for offering pirated digital goods. These two mechanisms imply that by uploading a given pirated product on a one-click hosting service, a single user increases his or her chances of obtaining other products in the future, by improving technical access and by facilitating the search process.

Profitability

In the monetary part of the market, potentially high profitability is the primary product-specific feature that drives digital piracy. In the case of digital products, fixed costs of reproduction are relatively low and the marginal costs are virtually negligible. Low costs of reproduction imply relatively high potential total profitability of those instances of digital piracy when a price can be extracted from the consumer, and could translate into high incentives for infringing parties to enter the market.

The total profitability depends on whether the infringer is able to charge a price for supply of pirated products and on the total size of the relevant market. As outlined in the previous chapter, this could happen in some cases, especially if suitable costless pirated alternatives are not available and the total cost of acquisition of a legitimate product is perceived as being too high. Such limitations can occur, for example, if a given digital good cannot be found on the non-monetary market (*e.g.* a live sport event) or if the quality is suspect or if the distribution of the legitimate good is geographically limited (*e.g.* a TV programme). In these cases, access to costless pirated products might be limited and consumers might be willing to pay a positive price for the pirated digital product.

Some suppliers of free pirated content are also able to generate substantial income from their illegal activities – even if not by way of direct payments from recipients of pirated material, but from other indirect revenue generation; for example by advertising revenue on websites or revenue generated from sponsored links. However this type of revenue can be generated only by a limited number of infringers (*e.g.* webmasters of infringing web pages) compared to the exponential number of infringing end users. Industry members have noted that while the number of infringers able to generate this type of revenue is limited, it poses a substantial threat of harm to rights holders including rights holders who have attempted to explore advertising driven distribution models.

Market scope

Regarding the size of the relevant market for paid pirated digital products, the wide access to the Internet makes the scope of the market for pirated digital products global, where the exchange is free of any geographical controls, or the limitations that apply to the exchange of physical goods (*e.g.* custom controls or trade barriers) or restrictions in the markets of legitimate digital products (*e.g.* geographical restriction of availability of a given product).

The global market scope strengthens the above presented incentives, by offering a prospect of a large number of potential customers which in turn increases the expectations of all types of possible non-monetary and monetary rewards.

Market exploitation (technology)

Numerous technologies facilitate the processes of reproduction and distribution of digital products, thus permitting the efficient exploitation of the markets. Hardware dependence implies that digital products can be replicated, consumed and stored only on a hardware device, usually a personal computer but increasingly on mobile phones and other devices. Thus the aggregated importance of this factor depends closely on the growth of the market for personal computers. As indicated by the BSA-IDC study one of the main reasons for recent growth of piracy, was the "rapid influx of new PC users [among] consumers and small businesses" (BSA-IDC, 2007).

Technological developments affecting reproduction

The technical feasibility of reproduction is a necessary condition for the supply of pirated digital content. In most cases technology permits the costless and instantaneous replication of digital products, which act as a driver that facilitates digital piracy. For most digital products stored as a computer file, it usually takes a few moments to duplicate it. However, easy and costless reproduction does not necessarily imply that a pirated copy of a product will be technically ready for further use, as in some cases a given copied digital product may not be interoperable with hardware available to a consumer.[4]

In some cases there are additional technical requirements necessary to circumvent mechanisms added by rights holders to prevent the unauthorised reproduction of digital content. These mechanisms introduce certain fixed monetary or non monetary costs to break the technological protection of a digital product. However, once the protection is broken the digital content can be easily copied essentially without most of those costs; though legal and moral transaction costs would still apply.

Examples of such protection include cases of a digital content stored on a copy-protected driver (*e.g.* Blue-ray Disc). Other examples would be transmitted or streamed content (television, movies, music, radio), particularly those incorporating copy-protecting encryption (*e.g.* some live streaming services use encryption against unauthorised downloading and copying of the streamed material).

Technological developments affecting distribution

Recent technological developments have significantly improved the distribution possibilities of digital goods, which also affect the distribution of pirated digital products. There are two main areas where these improvements concern the distribution process of digital goods: the improving efficiency of distribution methods and the broadening of Internet penetration and access.

The improving efficiency of distribution methods refers to the growing number of channels through which the pirated content can be distributed (discussed in Chapter 1). Most of these technological solutions (*e.g.* P2P networks) were developed for legitimate purposes and greatly contribute to technological innovation, although some specific solutions were designed with the particular purpose of digital piracy in mind (*e.g.* hacking software).

Box 2.1. Examples of technological solutions that limit supply of pirated content

A good example of a technological solution limiting supply of pirated digital content relates to sites that host "user-generated content" (UGC). Only a short time ago, many of these sites were rife with illegal content, and the only remedy for rights holders was to continuously search sites for unauthorised content, then notify the sites to take down that content. This remedy was unsatisfactory to the rights holders, as the content could be reposted much more quickly than it could be identified and removed.

In October 2007, many of the major UGC sites and content holders signed the "UGC Principles" (*www.ugcprinciples.com*). Pursuant to those principles, the UGC signatories agreed to, among other things, "implement state of the art filtering technology with the goal to eliminate infringing content on UGC services, including blocking infringing uploads before they are made available to the public." Other UGC sites have since implemented content recognition technologies, even though they have not signed the principles. (The UGC Principles also provide that, as an alternative, "UGC Services may, at their option, utilise manual (human) review of all user-uploaded audio and video content in lieu of, or in addition to, use of Identification Technology, if feasible and if such review is as effective as Identification Technology in achieving the goal of eliminating infringing content.")

The content recognition technologies referred to in the UGC Principles operate by creating "fingerprints" of a wide range of copyrighted audio and visual content. When a user tries to upload content to a UGC site, the content recognition mechanism automatically checks the content's fingerprint against the database, and, if it finds a match, takes appropriate action. This content recognition technology appears to be effective to some degree in practice, greatly reducing the amount of infringing content available on compliant UGC sites. However, some commentators signal certain problems related to the low rate of supply of digital identifiers by content suppliers that can negatively affect the functioning of these sites. Nevertheless, this type of content recognition technology can also be applied in other contexts, and shows great promise as a technological method to reduce digital piracy.

The advent of the Internet and the rapid increase of personal computers have resulted in growing access to information worldwide. However this also means that more and more consumers have access to pirated digital goods. The growth of Internet coverage was paralleled by the growth of possible technological solutions that enabled distribution of pirated content including P2P networks or one-click hosting services (discussed earlier).

While technology can be a driver of digital piracy, it can and should also serve as a negative driver, making less illegal content easily available. For example content recognition technology has been employed to prevent certain uses of works. Technological solutions aimed at reducing the supply of pirated digital content include screening and filtering technologies (Box 2.1). Such technologies may be more appropriate in some situations than others. For example, whilst many user-generated content (UGC)[5] websites have adopted content recognition technology, legislation in many countries, makes it clear that intermediaries should not be forced to generally monitor all content passing over their networks (OECD, 2007).[6]

Market risks

Several factors determine the risk involved in digital piracy, including the degree of social control and the quality of institutional infrastructure and the legal and enforcement risks. Remarkably, these factors are in most cases common for both supply and demand sides of markets. This is due to two reasons at least. First, in many cases agents that act in the markets for pirated digital products simultaneously demand and supply those products, so that the related market risks affect both types of behaviour. Second, as noted before, the term "digital piracy" is a not a universally defined legal concept so that existing legal solutions (and related market risks) in many cases do not apply to supply or to demand uniquely.

Legal and regulatory framework

The legal and regulatory frameworks aimed at preventing digital piracy are key local factors, as they affect the behaviour of main market actors. Legal systems grant suppliers of digital products with copyright protection, and provide legal instruments to protect those rights. These instruments allow copyright holders to take legal action against infringing parties and to claim compensation for potential losses. Hence, a strong legal framework, with effective enforcement, including through cross-border co-operation, can reduce both pirate supply and demand, while a weak one could be viewed as tolerant towards these activities. This hypothesis seems to be confirmed by the empirical study by van Kranenburg and Hogenbirk (2005) who found that economies with strong copyright protection regimes tended to report lower rates of piracy.

Legal frameworks can affect the levels of digital piracy, but only to the extent that nominal laws are enforced in practice and taken advantage of by copyright owners. If the resources devoted to enforcement are inadequate or intellectual property rights are not otherwise enforced by public authorities, or rights holders choose not to take action against copyright infringers, then the value of the nominal laws and regulations to the rights holders is low. Even the nominally strictest law could therefore potentially have no impact on the scale of digital piracy.

In addition, the risk of discovery and subsequent legal action must also be considered in relation to the potential consequences that infringers may face, in terms of the amount of any expected monetary penalty or the likelihood and duration of possible jail sentences. For instance, if the expected consequences are small, then even a high risk of discovery may have little practical impact on counterfeiting and piracy activity. Penalties reflect the consequences that infringing parties expect to suffer if they are discovered. In this regard, the impact of enforcement on the volume of digital piracy may vary considerably according to the severity of the sanctions. There is a significant difference, for instance, between a minor fine, which can effectively be absorbed compared to the cost of legitimate good, and a criminal prosecution that eventually results in a prison sentence. In fact, according to a study by BSA-Harris (2007), potential legal consequences could discourage some individuals from exchanging pirated digital goods.

Several authors studied the potential negative effect of strict institutional regulation on digital piracy, and the results mostly confirm the deterrent effect in the case of software piracy (*e.g.* Gopal and Sanders, 1997; Higgins and Makin, 2004a, 2004b; Higgins, 2005) and music piracy (Wade, 2004; Fivelsdal, 2005). The potential deterring effect of a strong legal system was also confirmed by survey based studies on consumer behaviour by IFPI (2008) and Ipsos (2008).

However, in general a risk of penalisation should be put in the context of total efficiency of the legal system. The key is not only the availability of remedies but the actual application of them in individual cases – and publicising that fact to others engaged in similar conduct. This is particularly important in cases where there is also a high degree of corruption, as this significantly reduces the efficiency of legal mechanisms to which industry can avail itself. Numerous empirical studies (Ronkainen and Guerrero-Cusumano, 2001; Papadopoulos, 2003; Bagchi *et al.*, 2006) found the degree of corruption to be positively correlated with the levels of digital piracy.

Parental control

Apart from weak and non-deterrent legislative solutions, the lack of parental control seems to be another factor that reduces the risk related to supply of pirated digital products. Lack of parental supervision seems to be an important part of the social environment for digital piracy, given the large number of youths operating in markets for pirated digital products. According to BSA-Harris (2007), parental control could act as a deterrent factor to prevent minors exchanging pirated digital products. As shown by this study, parents have a strong influence on their children's behaviour, and there is a clear difference between the activities of youth with and without rules and parental supervision.

Demand drivers

Factors that drive the demand for pirated digital products can be grouped into three sets (product-specific, individual and market specific) and can be derived from the market analysis on consumers decisions presented in the previous section.

As discussed in Chapter 1, there are three general conditions that need to be simultaneously fulfilled for a consumer to demand a given digital product (pirated or legitimate). First, a consumer needs to demand a given product in general. He or she must like the product so that the utility derived from its consumption offsets the price paid (if any) and the associated transaction costs. Second, a given product must be affordable and not violate the consumer's budget constraints. Third, a given product needs to be less expensive in terms of total costs (price and transaction costs) than other alternatives for a consumer to acquire it.

Thus, consumer decisions are affected by the overall utility a consumer derives from consumption of a product, the consumer's budget constraint and the transaction costs related to acquisition of a given product. The elements that determine the overall utility that a consumer expects to derive from a given product are product content and its technical quality. Factors that refer to a consumer's budget constraints include the product–specific price and the individual economic situation of the consumer. Lastly, factors that shape transaction costs incurred during the acquisition of a given product include:

- possible limitations of product's storage and consumption;
- some individual factors, such as security awareness and attitudes towards digital piracy;

- a set of market-specific factors that determine the overall availability of a pirated product; and
- the risks related to its acquisition and consumption.

In this context it is important to highlight that the new business models employed by suppliers of legitimate products (see Box 2.3) can play a significant role in reducing the incentives to demand pirated content. Table 2.2 summarises the product-specific, individual and market-specific demand drivers.

Table 2.2. Demand drivers of digital piracy

Driving factors	Characteristics that favour digital piracy
Product-specific	
Content	Attractive content
Technical quality	High technical quality or a copy
Consumption and storage	Easy consumption and storage
Price	Low or no price
Individual-specific	
Personal economic situation	Low disposable income (in case of young market participants)
Security awareness	Low awareness of potential malicious software risks
Attitude towards piracy	Permissive attitude towards digital piracy
Market-specific	
Product availability	Large availability of pirated digital content
Market risks	Inefficient legal infrastructure, lack of parental control

Product-specific drivers

What product-specific characteristics could drive demand for pirated digital piracy? First, the content of a digital product is the primary motive of its acquisition; in particular if it becomes very popular among consumers. Second, a copy of a digital product can offer a relatively high quality, indeed often the same quality as the original good. Third, a pirated copy can be easily consumed and stored. Last, the pirated digital products are offered at a low or even no price.

Content

Generally speaking, the higher the attractiveness of the content of a given digital product, such as music, movies, TV programs or software, the higher the likely demand so that high interest in content is one of the key product-specific features that increases the probability of an infringement.

In many cases the content itself seems to be more important for consumers than the technical quality of a particular copy. In the case of music and movie products, the content attractiveness can be strengthened by the visibility of artists and strong followings. Thus, a given digital product can evoke consumer expectations with respect to attractiveness of the product and the image projected by an artist. An example of this is the so-called "pre-release piracy", which occurs before the official release date of a digital product.[7] Another example is the broadcasting of television programs, such as live sport events, which are particularly attractive, given their relatively high depreciation rate, so that any delay results in a significant drop of utility. The case study on the sports broadcasting rights industry illustrates this point. The attractiveness of the content of entertainment software products can also be strengthened by its popularity and visibility.

The importance of content in consumers' decision making processes is reflected in empirical studies whether of digital or other products. According to the report by Bryce and Rutter (2005), for most consumers quick access to a given digital content, such as music or movies, is a major motivation for acquiring a pirated copy, which seems to indicate that the emergence of legal business models offering quick access to content may diminish the attractiveness of piracy.[8]

The utility that individuals derive from the content of a digital product could be expected to be closely linked to its technical quality. However, whereas product quality describes features specific to a particular copy of a digital product, the content reflects the recognition a product enjoys among the consumers. Thus, several digital products with the same content (*e.g.* several different files containing a given music title) can have different technical quality, and each could be acceptable to consumers according to the circumstances in which it is acquired.

Technical quality

Technical quality of a given product includes other-than-content aspects of its total attractiveness and refers to its technical specifications and related usefulness. Higher technical quality are factors (although not the main factors) driving demand for a given pirated digital product.

For example, the technical quality of a music product will mostly reflect the quality of the sound (expressed, for example, as bit-conversion rate) and compatibility with various music players. In the case of a movie product, they can be captured by the quality of image and sound as well as the possibility of playing on various movie players. For software they will reflect its reliability, mobility, etc.

In many cases, pirated copies of a digital product can offer virtually the same quality as the genuine good, which means that from the perspective of quality a consumer is likely to be indifferent between genuine and infringing goods. This is likely to happen if a pirated product is acquired from a local source or from a source of a high degree of trust (such as local Internet community exchanging pirated digital products via one-click hosting services).

However, in some cases pirated copies of digital products offer significantly lower quality than the legitimate products. Pirated music files are reported to offer worse quality of sound (IFPI, 2006), while movies are generally found to have poorer video and sound quality.[9] Pirated pre-release versions of video games may have numerous software bugs that would not appear in the final, published version. Streaming services offering pirated video or television products are of substantially lower quality in terms of image resolution and sound quality than those that are legitimates, although there are indications that, especially in the case of pirated sport broadcasts, the technical quality is improving, and the difference is narrowing. Pirated files are sometimes found to contain malicious software and viruses.

Even though technical quality of some copies can be lower than the quality of legitimate products, in certain cases, (where the content and/or timeliness is of paramount importance), the technical quality and functionality seem to be less important. According to the study of Bryce and Rutter (2005) consumers tend to place higher emphasis on their being able to access the digital content, rather than being assured of its high technical quality.

Consumption and storage

Once acquired, a pirated digital product must be consumed and sometimes stored using a given piece of hardware in order to fulfil the user's expectations. The technical ability to consume and to store a given digital product is another technical feature that drives digital piracy.

Hardware dependence implies that digital products can be replicated and consumed and stored only on a hardware device, usually a personal computer. Thus the aggregated importance of this factor depends closely on the growth of the market for personal computers. As indicated by the BSA study one of the main reasons for recent growth of piracy, was the "rapid influx of new PC users in the consumer and small-business sectors" (BSA, 2007).

Numerous computer programs, devices and hardware equipment available on the market facilitate the consumption and storage of digital products. This includes a large variety of computer media players and readers that often are delivered together with the computer or can be downloaded for free from the Internet. The strong market hardware designed for media consumption, such as portable music players or PDAs could also affect the degree of digital piracy. A study by Ethier (2007) indicated that the market for portable music players enjoys high growth rates.

The large increases in recent years in storage capacities, including web servers, volumes and numbers of local hard drives, could also be perceived as factors driving digital piracy. This growth in volume was complemented by an observable reduction in the cost of storage media such as hard discs and portable devices including optical discs and flash drives. As indicated in numerous studies on digital content (OECD, 2005 and 2007), a growth in storage capacities act as a driver for supply and storage of different digital content, hence it is possible that it can act as a driver of digital piracy as well.

Box 2.2. Digital rights management (DRM)

DRM refers to technologies used by owners of copyright and related rights to facilitate rights clearance, inform consumers of permitted uses, and prevent unauthorised usage of their protected content. The DRM is fundament of a new business utilised by rights holders to prevent loss of their profits (Varian, 2005).

In practice, DRM technologies consist of two aspects. "Right management information" provides consumers with information regarding permitted uses and copyright ownership, facilitating use and clearance of rights. "Technological protection measures" attempt to prevent unauthorised use of protected content by limiting access, copying or other unauthorised actions by end users conditional on compliance with licensing conditions applied by the owner of rights. rights managements also includes metadata that identify specific digital content, the authors or rights holders, and express licensing terms for use of the content.

According to a study published by the European Commission, (European Commission, 2006) these DRM technologies are legally implemented in most national laws of the EU and the unauthorised circumvention or removal of DRM protection is unlawful. The importance of policy efforts to improve digital rights management and the development of new, attractive business models as tools to reduce piracy have already been highlighted in an earlier OECD study on "Digital Broadband Content: Digital Content Strategies and Policies" (OECD, 2006b).

Box 2.3. New legal delivery channels for digital products

The market for legitimate digital products has evolved significantly in recent years. New business models rely on the Internet as a means to promote and distribute digital products. Internet enables new businesses not only to sustain themselves by generating profits, but also to compete successfully in a rapidly evolving market environment.

In the context of digital piracy, the new solutions have significantly improved the overall attractiveness of legitimate digital products and hence (at least partially) reduced the potential incentives to engage in illegal markets. It has been achieved by the dramatic improvement of the searching process and immediate delivery of a product (which implies a significant reduction in transaction costs). The customer also has much more flexibility in selecting the final product, and prices are usually lower than prices of comparable products offered through traditional distribution channels.

Examples of these business solutions include *iTunes*, *Play.com* or *last.fm.*

One of the purposes of the technology known as digital rights management (DRM; see Box 2.2) by rights holders is to reduce the unauthorised use of digital content and enable innovative business models (see Box 2.3). The principle of digital rights management is to prevent or condition access to the content through technical means such as encryption, or, more recently, use of digital fingerprinting and watermarking technologies. In the context of digital piracy this solution is a way of controlling the use of content by reducing the possibility of those digital products being consumed by those who are not entitled to do so. Therefore DRM can be classified as a potential negative driver of digital piracy. On the other hand some studies indicate that DRM could reduce the degree of technical availability and mobility of digital products (see OECD, 2006a) and therefore indirectly act as a positive driver of digital piracy (as discussed later in the subsection on product availability). Thus, the total effect of DRM on digital piracy seems to be ambiguous.

Price

The low or zero price of pirated digital products is one of the key drivers of digital piracy. Generally price is one of the main elements that influence consumer decisions on whether or not to acquire a given good, and not unexpectedly the price of pirated digital products relative to the price of the legitimate counterparts is a very significant influence. Indeed as reported in the study by Bryce and Rutter (2005), a low or zero price was the most frequently cited motivation of acquisition of pirated digital products.

Low or no price for pirated digital products is perceived by the consumers in the context of comparatively higher prices of legitimate products. As for other economic exchanges, there exists a price equilibrium at which consumers might be prepared to pay suppliers what they see as a fair price for a (digital) product. Therefore, taking advantage of the cost saving innovations enabled by digital technologies could help legal suppliers reach that price equilibrium. However, the market distortions caused by digital piracy should not be underestimated since the infringing parties do not bear all the fixed costs related to the creation of content and other costs. Therefore, legitimate suppliers will never be able to offer their products/services on a level playing field *vis-à-vis* pirate sites that do not incur these costs. In this regard, new legal delivery channels cannot be considered as a universal and unique solution to counter digital piracy, even though they seem to reduce significantly the incentives for the demand of digital products.

The recent success of some new and legal delivery channels for digital products would reinforce this case, whereby not only would piracy be countered, but the emergence of (legal) music downloads seems to have also led to increases in consumption patterns (Box 2.3).

Individual-specific drivers

There are at least three types of demand drivers of digital piracy that are specific to the individuals involved. The first concerns the income of individuals and the general economic situation which is related to a consumer's budget constraint. The other two include some of the transaction costs involved in the acquisition of a pirated digital product including: potential risk related to malicious software threat and psychological disutility that comes from individual attitudes towards digital piracy.

Personal economic situation

Apart from the price, the personal economic situation of individuals determines their budget constraints, *i.e.* limits of their ability to purchase goods. In some instances, disposable income levels could be a decisive factor in a consumer's decision to acquire a pirated digital product, which may seem like a "free" alternative to legitimate products.

Some empirical studies tend to support this hypothesis. According to Bryce and Rutter (2005) personal economic situations and lack of access to credit cards, was perceived to be a barrier to many young consumers accessing legal downloading services. Cheng *et al.* (1997) identified the relationship between software price, household income and software piracy rate as statistically significant. They suggested that a higher software price might increase consumers' willingness to buy pirated copies. However, there are no studies that could contribute to the robustness of these results or that would generalise them for other industries that supply digital products.

On a macroeconomic, economy-wide level, low personal income would imply, all things being equal, that poor countries should have greater piracy rates than wealthy ones, and several studies have attempted to identify a relationship between aggregated piracy level and per capita income at the economy level. The main finding in most studies is that economic factors are significant on their own, but their explanatory power shrinks when other factors (such as cultural or legal) are included in the exercise. Some studies suggest (*e.g.* Gopal and Sanders; 1998, 2000) that real GDP per capita has an impact on the piracy rate in economies with lower levels of income than in wealthier ones, which suggests that the positive impact of low personal income on digital piracy is particularly strong in low income economies.

Lack of security awareness

Security awareness obviously figures prominently in individuals' decisions on whether or not to engage in digital piracy. In most cases a potential security threat discourage aware users from engaging in digital piracy,[10] hence as a corollary the lack of security awareness can be considered as a positive driver for digital piracy.

Most of the concerns about security risks involve the potential exposure to malicious software, which is designed to infiltrate or damage a computer system without the owner's knowledge or consent. Malicious software includes computer viruses, worms, trojans, spyware, dishonest adware, and other malicious and unwanted software (OECD, 2008). According to a study by BSA-Harris (2007) the threat of viruses and problems with spyware are the biggest potential safety risks associated with the acquisition of pirated digital products. Many sources of digital pirated content such as warez sites or misused P2P networks are reported to be the source of malware, and individuals that engage in digital piracy using these sources may be exposed to this risk.

Some participants in the pirated digital products market recognise these risks. According to Bryce and Rutter (2005), some users identify computer viruses as one of the main risks related to digital piracy. The results of a survey presented by Wolfe *et al.* (2008) reveal that fear of computer viruses may influence respondents' intentions to engage in digital piracy. However, a large volume of users does not even seem to be aware of potential security risks, and even those that some knowledge of potential security threats do not consider these threats to be significant. According to an empirical study by IFPI (2006), only 35% of respondents reduced their music piracy activity as a consequence of the effects of viruses on their computer.

Attitude towards digital piracy

A perception that acquisition of pirated digital product is illicit and blameworthy can potentially create a moral disutility to a potential consumer and as a consequence become a significant transaction cost and negative demand driver. However, as outlined before, many users do not perceive digital piracy as morally unacceptable, especially in cases when no monetary payment is involved.[11]

The permissive attitude in case of demand for pirated product seems to be even stronger than in the case of pirate supply. As reported by a number of other studies (Solutions, 2006; BSA-IDC, 2007) the consumption of pirated digital content nay be perceived as "normal", particularly among younger users. In addition, some consumers are reluctant to accept digital piracy as "theft", especially when no monetary profit is perceived to be generated by the supplying parties.

In fact, according to numerous studies, consumers treat digital copyright infringements differently from other infringements of the law. Downloading of a pirated movie for example, might not be considered in the same way as the petty theft of a small amount of money, even though both acts might result in a similar monetary loss for someone. According to the report by Bryce and Rutter (2005) consumers of pirated digital products are mostly aware that digital piracy is illegal. However, since they mostly acquire the pirated digital products free of charge, this is not perceived as problematic ethically.

Market-specific drivers

Numerous market-specific factors drive demand for pirated digital products. The term market-specific refers to factors that affect transaction costs on markets where the exchanges of digital products are made. These factors include the availability of digital products and market risks related to their consumption.

Product availability

Wide access to the Internet makes the scope of the market for pirated digital products global which through the lack of practically any geographical obstacles becomes a driving factor of digital piracy. Regulatory barriers to multi-territory licensing of digital content and products may also enhance the desirability of pirated products due to the lack of availability of a legal alternative.[12]

The global scope of the market for pirated digital products becomes important when compared to markets for legitimate digital products. In many cases the market scope of a legitimate product is geographically limited and certain cross borders transactions are not allowed in certain markets. In these cases the global scope of the market for pirated digital products gains additional importance, since it offers potential access to a legally unavailable product. This effect was particularly noted in the case study presented in Chapter 4, where contractual arrangements frequently prevent live broadcasts of sporting events from being legally available in certain countries, territories or regions.

Another dimension of the global scope of the markets for pirated digital goods is the recent technological developments (the broadening of Internet access and the improving efficiency of distribution methods) that have significantly improved the distribution possibilities of digital goods, including those that are pirated.

Other technological developments affect the efficiency of distribution through improved searching processes and innovations that augment the efficiency of transfer of digital products. These include the rapid growth of search engines and the emergence of various Internet services that can be used to exchange pirated digital content (presented in Chapter 1). Even though these services are not normally devoted to the exchange of pirated goods, they may significantly facilitate the exchange.

Market risks

Even though digital piracy is a phenomenon of global scope, its intensity tends to vary across different economies. One of the potential explanations for these cross-economy differences is the presence of sets of local, economy-specific factors such as legal and regulatory frameworks or cultural and social characteristics, which affect the risks related to digital piracy.

These factors include the degree of parental control and the quality of legal infrastructure. Remarkably, for a variety of reasons, these factors are common for both the market supply and demand sides (discussed earlier).

Some elements of the total legal infrastructure seem to be more demand-specific, however. Increased media coverage of piracy and the rights holders' anti-piracy efforts have also increased the perceived risks faced by consumers engaged in exchanging pirated products (see Box 2.4 for specific examples). This increases consumer awareness of potential legal problems related to consumption of pirated digital products, although a majority of consumers seem to believe that the likelihood of being caught is very small (Wiggin, 2008).

Box 2.4. Legal actions to reduce demand for pirated digital products

Examples of proposed solutions to counter digital piracy by reducing demand include French "three-strikes-and-you're-out" memorandum of understanding (MOU), the agreement between Virgin and the BPI in the United Kingdom, and other actions.

In 2008, a cross-industry MOU was signed in France for a scheme where anyone who persists in illicit downloading of music or films will, after a series of warnings, be barred from broadband access (also known as Olivennes Agreement). The scheme is currently a proposed draft bill (the Creation and Internet Bill), and has not yet been implemented. The proposal includes administration of the scheme by an independent government body under the supervision of a judge.

In the United Kingdom, one of the ISPs sent about 80 letters to customers warning them that they should not be downloading illegal music files via file-sharing sites. This action was undertaken in co-operation with the BPI (formerly known as British Phonographic Industry), the body that represents the British recorded music business and was part of an educational awareness campaign. The letters were educational in tone, informing customers of the implications of sharing copyright works, and did not threaten disconnection of customer accounts.

Similarly, in other countries (*e.g.* Australia, Japan, Korea and New Zealand) rights holders have proposed various legal options that foresee the involvement of Internet Service Providers in combating digital piracy through a graduated suspension or disconnection procedure of those involved in infringement.

It is acknowledged that these types of measures are not universally accepted. The controversies related to this type of demand reducing remedies were expressed by the resolution of EU Parliament which called on its Member States, to "… avoid adopting measures conflicting with civil liberties and human rights and with the principles of proportionality, effectiveness and dissuasiveness, such as the interruption of Internet access." (EU Parliament Resolution, 10 April 2008).

Notes

1. In social sciences the term "collectivism" (as opposed to "individualism") refers to a setting in which the collective is more important than the separate individuals and "…the view that the whole is greater than the sum of its parts" (Agassi, 1960).

2. Studies that used Hofstede's model in analysis of digital piracy include: Ronkainen and Guerrero-Cusumano (2001), Kyper *et al.* (2004), Depken and Simmons (2004), Bagchi *et al.* (2006), and Ki *et al.* (2006).

3. Power distance index proxies the extent to which a society's level of inequality of power is supported by the less powerful individuals as much as by the most powerful ones.

4. This aspect is presented in the section that studies the demand drivers – technological requirements with respect to consumption of digital products.

5. User-Generated Content (UGC) is also known as User-Created Content (UCC), see OECD (2007).

6. European E-commerce Directive (2000/31/EC)

   Article 15 No general obligation to monitor

   1\. Member States shall not impose a general obligation on providers, when providing the services covered by Articles 12, 13 and 14, to monitor the information which they transmit or store, nor a general obligation actively to seek facts or circumstances indicating illegal activity.

   2\. Member States may establish obligations for information society service providers promptly to inform the competent public authorities of alleged illegal activities undertaken or information provided by recipients of their service or obligations to communicate to the competent authorities, at their request, information enabling the identification of recipients of their service with whom they have storage agreements.

7. As indicated by the industry, pre-release piracy is one of the most economically damaging form of piracy, denying artists revenue at the most crucial period in the lifecycle of a creative work.

8. Examples of such legal content delivery services include *iTunes*, *Amazon*'s music download system, or *play.com*, to name but a few.

9. This is especially the case if a copy was illegally recorded in the cinema.

10. Thus in the broad economic context it can be considered as a transaction cost.

11. See discussion on supply drivers, section on "attitudes towards digital piracy".

12. As mentioned for instance in the 2008 European Commission consultation on creative content online, available at: *ec.europa.eu/avpolicy/other_actions/content_online/consultation_2008/index_en.htm*

References

Agassi, J. (1960), "Methodological Individualism", *British Journal of Sociology*, 11(3): 244-270.

Bagchi, K., P. Kirs and R. Cerveny (2006), "Global Software Piracy: Can Economic Factors Alone Explain the Trend?", *Communications of the ACM*, 49(6), 70-75.

Bryce, J. and Rutter, J. (2005), "Fake Nation?", available at *www.allianceagainstiptheft.co.uk.*

BSA - IDC (2007), "Fourth Annual BSA and IDC Global Software Piracy Study", Business Software Alliance, available at *www.bsa.org/globalstudy,* accessed 6 March 2008.

Cheng, H. K., R.R. Sims and H. Teegen (1997), "To Purchase or to Pirate Software: An Empirical Study", *Journal of Management Information Systems*, 13(4), 49-60.

Depken, C.A. and L.C. Simmons (2004), "Social Construct and the Propensity for Software Piracy", *Applied Economics Letters*, 11(2), 97-100.

EC (2006), "Interactive Content and Convergence: Implications for the Information Society", a study for the European Commission, DG Information Society and Media.

Ethier, S. (2007), "Worldwide Demand Remains Strong for MP3 and Portable Media Players", in *Stat*.

Fehr, E. and S. Gächter (2000), "Fairness and Retaliation: The Economics of Reciprocity", *Journal of Economic Perspectives*, 2000(14), 159-181.

Fehr, E., S. Gächter and S. Kirchsteiger (1997), "Reciprocity as a Contract Enforcement Device: Experimental Evidence", *Econometrica*, vol. 65, no. 4, July, pp. 833-860.

Fivelsdal, H. (2005), "Moving toward a Balanced and Effective Response to Internet Music Piracy", *The International Journal of Media Management*, 7.

Gopal, R.D. and G.L. Sanders (1997), "Preventive and Deterrent Controls for Software Piracy, *Journal of Management Information Systems*, 13.

Gopal, R.D. and G.L. Sanders (1998), "International Software Piracy: An Analysis of Key Issues and Impacts, *Information Systems Research*, 9(4), 380-397.

Gopal, R.D. and G.L. Sanders (2000), "Global Software Piracy: You Can't Get Blood out of a Turnip", *Communications of the ACM*, 43(9), 82-89.

Higgins, G.E. (2005), "Can Self-Control Theory Help Understand the Software Piracy Problem?", *Deviant Behavior*, 26.

Higgins, G. E. and D.A. Makin (2004a), Does Social Learning Theory Condition the Effects of Low Self-Control on College Students' Software Piracy? *Journal of Economic Crime Management*, 2.

Higgins, G. E. and D.A. Makin (2004b). Self-Control, Deviant Peers and Software Piracy, *Psychological Reports*, 95.

Hofstede, G. (1980), *Culture's Consequences: International Differences in Work-Related Values*, Beverly Hills, California: Sage Publications.

Hofstede, G. (2004), "Geert Hofstede Cultural Dimensions", *www.geert-hofstede.com/hofstede_dimensions.php*, accessed 15 April 2008.

IFPI (2006), "The Recording Industry 2006 Piracy Report: Protecting Creativity in Music", International Federation of Phonographic Industries.

IFPI (2008), "IFPI Recording Industry in Numbers", International Federation of the Phonographic Industry.

Ipsos (2008), "Les Français face au téléchargement illégal de musique sur Internet", survey by Ipsos and SCPP (Société Civile des Producteurs Phonographiques).

Ki, E.-J., E.-H. Chang and H. Khang (2006), "Exploring Influential Factors on Music Piracy across Countries". *Journal of Communication*, 56(2), 406-426.

van Kranenburg, H., and A. Hogenbirk (2005), "Multimedia, Entertainment and Businеss Software Copyright Piracy: A Cross-National Study, *Journal of Media Economics*, 18(2), 109-129.

Kyper, E., R.J. Lievano, P. Mangiameli and S.K. Shin (2004), "oftware Piracy: A Time Series Analysis", Proceedings of the 10th Americas Conference on Information Systems, New York.

Lenhart, A.M, M., A. Rankin Macgill and A. Smith, (2007), "Teens and Social Media: The Use of Social Media Gains a Greater Foothold in Teen Life as Email Continues to Lose its Luster", Washington, DC: Pew Internet & American Life Project, 19 December.

OECD (2005), "Digital Broadband Content: Music", DSTI/ICCP/IE(2004)13/FINAL, available at *www.oecd.org/sti/digitalcontent,* OECD, Paris.

OECD (2006a), "Disclosure Issues Related to the Use of Copy Control and Digital Rights Management Technologies", OECD, Paris, DSTI/CP(2005)15/FINAL.

OECD (2006b), Digital Broadband Content, Digital content strategies and policies, Working Party on the Information Economy, DSTI/ICCP/IE(2005)3/FINAL.

OECD (2007), *Participative Web and User-Created Content: Web 2.0, Wikis and Social Networking,* OECD, Paris.

OECD (2008), *Computer Viruses and Other Malicious Software: A Threat to the Internet Economy*, OECD, Paris.

Papadopoulos, T. (2003), "Determinants of International Sound Recording Piracy", *Economic Bulletin*, 6(10), 1-9.

Ronkainen, I.A. and J.L. Guerrero-Cusumano (2001), "Correlates of Intellectual Property Violation", *Multinational Business Review*, 9(1), 59-65.

Solutions (2006), "Movie File-Sharing Booming", Solutions Research Group, available at *www.srgnet.com*, accessed 28 March 2008.

Varian, H.R. (2005), "Copying and Copyright", *Journal of Economic Perspectives*, 19(2).

Wade, J. (2004), "The Music Industry's War on Piracy", *Risk Management*, 51.

Wiggin Digital Entertainment Survey (2008), available at *www.wiggin.co.uk,* accessed 17 April 2008.

Wolfe, S. E., G. E. Higgins and C.D. Marcum (2008), "Deterrence and Digital Piracy: A Preliminary Examination of the Role of Viruses", *Social Science Computer Review* 26.

Chapter 3

Industry Initiatives and Institutional Remedies

This chapter presents industry initiatives to counter digital piracy and institutional remedies for the problem.

Industry initiatives

The term "industry" is used here to denote copyrights holders and refers to two types of associations: *i)* cross-sector associations that are devoted to combating counterfeiting and piracy, including digital piracy, and *ii)* sector associations in a given industry that, among other issues, deal with the problem of digital piracy.

The main cross-sector industry initiatives of important international dimension were outlined in the Phase I report of the OECD's study on counterfeiting and piracy (OECD, 2008).[1] These initiatives are devoted to combating counterfeiting and piracy, including digital piracy.

Apart from the cross-sector initiatives, many intra-sector groups undertake actions to prevent and to combat digital piracy in their industry. Examples of such intra-sector initiatives are outlined in Table 3.1.

There are numerous activities taken by these initiatives that are related to different drivers of digital piracy (outlined in Chapter 2). Rights holders' initiatives focus on drivers that can be affected in a piracy-reducing way, including: security awareness, attitudes towards digital piracy, technological developments and legal and regulatory frameworks. A brief overview of drivers of digital piracy and corresponding industry actions is presented in Table 3.1.

To target these digital piracy drivers, rights holders undertake different actions in various areas including: *i)* collecting intelligence and undertaking educational campaigns; *ii)* co-operating and co-ordinating at the industry level; and *iii)* co-operating and supporting government efforts to combat digital piracy.

Table 3.1. Examples of sector initiatives that counter digital piracy

Organisation	Overview
BPI (formerly known as British Phonographic Industry)	The BPI represents the British recorded music business. In the context of digital piracy, BPI protects the interests of its members by enforcing copyright law in the United Kingdom. It does so through a variety of means, including civil litigation, criminal litigation, lobbying against IPR violating sites, and supporting other enforcement agencies in anti-copyright infringement raids. Membership: 400 members, large and small, who between them account for in excess of 90% of the UK music market. Website: *www.bpi.co.uk*
Business Software Alliance (BSA)	BSA brings together the world's largest commercial software producers and their hardware partners. BSA's principal activities focus on countering software piracy, through policy initiatives that promote copyright protection and education programmes. The BSA regularly publishes annual software piracy studies that monitor the markets for pirated software and aim at measurement of its scale. Membership: 31 large software producers. Website: *www.bsa.org*
Content Delivery and Storage Association, (CDSA)	Formerly IRMA (International Recording Media Association), CDSA established the Anti-Piracy Certification/Compliance Program for manufactured optical media (CDs, DVDs and CD-ROMs), analogue media (tape, vinyl) and other recording media formats. The goal of the program is to establish manufacturing procedures and systems that are designed to reduce the risk of pirated or unauthorised material being published. Membership: about 150 member companies in 23 economies. Website: *www.contentdeliveryandstorage.org*
Entertainment Software Association (ESA)	This US trade association represents publishers of entertainment software, such as video and computer games for video game consoles, personal computers, and the Internet. One of ESA's objectives is to combat piracy of global entertainment software through its Anti-Piracy Program. Its main components are policy work, training and education, and various enforcement efforts in the United States and abroad. Membership: 23 companies. Website: *www.theesa.com*

Table 3.1. Examples of sector initiatives that counter digital piracy *(continued)*

Organisation	Overview
International Federation of the Phonographic Industry (IFPI)	IFPI represents the interests of the recording industry worldwide. One of IFPI's mission statements is to "safeguard the rights of record producers". To do so, the organisation acts at global and regional levels. At the global level it provides a range of global industry statistics and co-ordinates international strategies in such areas as anti-piracy enforcement, technology, legal strategies, litigation, public relations, lobbying of governments and representation in international organisations. These strategies are then implemented at the regional level, together with co-ordination of the work of local groups, and setting local lobbying priorities. Membership: about 1 400 record companies in 73 economies and affiliated industry associations in 48 economies. Internet: *www.ifpi.org*
Motion Picture Association (MPA) / Motion Picture Association of America (MPAA)	These associations represent the American motion picture, home video and television industry, domestically through the MPAA and internationally through the MPA. They publish industry-specific research and statistics. They also counter piracy in the audiovisual industries through educational campaigns promoting copyrights, promoting technological solutions to combat digital piracy and co-operation with national and international policymaking agencies, working with anti-piracy programmes throughout the world. They work with local film industry to lobby for responsible content protection, effective enforcement and education campaigns. Membership: 6 biggest US movie producing companies. Website: *www.mpaa.org*
Music Publishers' Association of the United States (MPA)	MPA represents US producers and distributors of sheet music. It advocates the protection of intellectual property rights in the legal and legislative arena and leads educational campaigns promoting copyrights. Membership: 73 US music publishers. Website: *www.mpa.org*

Table 3.1. Examples of sector initiatives that counter digital piracy *(continued)*

Organisation	Overview
Recording Industry Association of America (RIAA)	RIAA groups the recording industry in the United States. One of its stated goals is "to protect intellectual property rights worldwide and the First Amendment rights of artists". RIAA conducts consumer, industry and technical research, and monitors and reviews state and federal laws, regulations and policies. Membership: 1 600 member labels (record labels and distributors). Website: *www.riaa.org*
Software and Information Industry Association (SIIA)	SIIA gathers US software and digital content producers. It runs a corporate anti-piracy program that identifies, investigates, and resolves software piracy cases on behalf of its members. Membership: about 750 US companies. Website: *www.siia.net*
U.S. National Music Publishers' Association (NMPA)	NMPA gathers US music publishing companies. It engages in work with government on legislative, litigation, and regulatory issues. Membership: over 700 U.S. companies. Website: *www.nmpa.org*
Sports Rights Owners Coalition (SROC)	This is an informal group of representatives of international and national sports bodies with a particular focus on broadcasting rights issues. It operates as a forum through which sports bodies can share information and experiences. One of purposes of SROC is to enable sports to take joint action to protect and promote their rights, including the prevention of piracy of their events. Membership: 38 sports associations worldwide.

Table 3.2. Digital piracy drivers and corresponding rights holders' action areas to counter digital piracy

Digital piracy driver	Action areas
Security awareness	Educational campaigns to raise awareness of potential security issues.
Attitude towards piracy	Educational campaigns highlighting the unethical character and potentially damaging effects of digital piracy.
Technological developments	Within-industry co-operation and co-ordination of the development and introduction of piracy-countering technologies.
Legal and regulatory framework	Co-operation with governments and agencies. Within industry, co-operation to promote copyright enforcement. Education of those involved in copyright enforcement.

Data collection and educational campaigns

Rights holders are an important source of information on digital piracy, including statistical data and research as well as educational initiatives. Data and economic analyses provided by the rights holders are essential to understand the scale of digital piracy and to identify the main areas of threat within each sector. Collected intelligence, together with input from other industry groups, can also act as the basis for further, more specific actions countering digital piracy.

Effectively reducing digital piracy requires a fundamental shift in the public's attitude toward this practice, and public education is a critical component of any successful effort. Educational campaigns seem to be particularly important, as they build social awareness of digital piracy, and the lack of awareness of the threat of digital piracy and its potential negative consequences has been identified as one of the key driving motives of copyrights infringement, which points at the long term importance of such actions. Three types of education-raising campaigns can be distinguished: *i*) those that raise the awareness of potential security risks; *ii*) those that are devoted to changing the attitude towards digital piracy; and *iii*) those that are focused on individuals involved in enforcement of the digital piracy-related regulations.

There are campaigns that are focused on potential security risks, based on security awareness as a driver of digital piracy. These campaigns include the provision of general information on this type of threat (*e.g.* BSA's Cybersafety) including screening services devoted to detecting potential safety threats (*e.g.* McAfee Site Advisor).

Other educational campaigns focus on the generally permissive attitude towards digital piracy and stress the unethical aspect of this procedure and its potentially damaging consequences. Many of the organisations mentioned above participate in such public awareness programmes. The World Intellectual Property Organization (WIPO), for example, conducts various awareness-raising activities, and the BSA has launched awareness programmes such as "Define the Line". The ESA has an awareness programme for students in grades K-5 called "Join the © Team", which provides educators with age-appropriate materials on basic copyright principles. In 2008 Pro-music and Childnet International launched the global education campaign "Young People, Music and the Internet," which was supported by the European Commission and several local governments.

Another example is the movie industry that has funded or supported educational campaigns in a number of territories throughout Europe. Some of these initiatives have been supported alongside other rights holders' groups (*e.g.* music and software) including campaigns in Finland, Hungary, Italy and Poland. In other countries the MPA along with the local industry has funded campaigns focusing specifically on the film industry and the impact piracy has on future creativity. These include the three-year strategy launched by Film Education in the United Kingdom, the *RespectCopyright.de* website in Germany and the "Tour of France" road show visiting schools in France organised by the educationalists Calysto. Creative campaigns, inviting students to develop anti-piracy campaigns have recently been successfully launched in the United Kingdom and Greece.

Some campaigns are directed to particular audiences; for example, Web Wise Kids[2] is a programme aimed at educating students about Internet safety issues.

Other examples of programmes aimed at public awareness in a particular economy are:

- In India, the Scheme of Intellectual Property Education, Research and Public Outreach and the National Initiative against Piracy and Counterfeiting formed by the Federation of Indian Chambers of Commerce.[3]
- In Australia, joint initiatives are conducted by the Australian Federal Police with organisations such as the Music Industry Piracy Investigations (MIPI); an example of this is the "Music for Free?" campaign.
- In 2006, in Hong Kong, China, the Customs and Excise Department launched the Youth Ambassador against Internet Piracy Scheme, with the objective of working with members of youth organisations

to serve as "web police" and report on suspected copyright infringements on the Internet ("Youth Ambassador", 2008).

- In Mexico, the campaign "Children against Piracy" was launched for the first time in 2007. It targets children between 6 and 12 years old and its main objective is to make children aware of the creativity and innovation value of IPR enforcement. Another campaign, "The Change Starts with an Idea ... It Can Be Yours", is addressed to people aged between 12 and 35 and its main objective is to promote the legal use of technological tools and to recognise the value of creativity, innovation and IPR enforcement.

The last aspect of education campaigns is the education of those involved in copyright enforcement, which is distinct from raising awareness of copyright issues in the wider community. Many of these training initiatives have been initiated in the United States. In 2006, for example, the US Patent and Trademark Office (USPTO) established the Global Intellectual Property Academy, which provides training for foreign officials on IP policy, best practices and enforcement. The United States has also posted eight "IP attachés" in six countries – China, Russia, India, Thailand, Egypt and Brazil – to work with the USPTO, United States embassies and local embassy task forces (NIPLECC, 2008).

The United States has also stationed IP Law Enforcement Coordinators in its embassies in Bulgaria and Thailand. In the United States, the National Intellectual Property Law Enforcement Coordination Council (NIPLECC) has a close partnership with several US and international trade organisations such as the BSA, IFPI and the Coalition Against Counterfeiting and Piracy (an initiative between the US Chamber of Commerce and the National Association of Manufacturers). Co-operative initiatives include the "Supply Chain Tool Kit" and industry conferences for those affected by IP crime, particularly software and film companies (NIPLECC, 2008).

A major initiative undertaken by the Office of the United States Trade Representative (USTR) in this area is the Special 301 Report. Pursuant to Special 301 provisions of the Trade Act 1974, the USTR examines the IPR protection regime and enforcement practices of its various trading partners throughout the world and designates countries in the categories of Priority Watch List, Watch List, and/or Section 306 Monitoring Status. The interagency Trade Policy Staff Committee co-ordinates through USTR on the issuance of the Special 301 Report, obtaining information from the private sector, US embassies, foreign governments and the US Congress "among other sources" as to the state of IPR protection and enforcement in the various countries identified in the report. With regard to digital piracy, the 2008 report identifies several notorious "virtual markets" through which pirated content is made available on line. It also notes the importance of implementing WIPO Internet Treaties, which should serve as the baseline for improving standards designed to address the global problem of Internet or digital piracy. Given that in many countries the government remains the primary vendee for copyrighted content, in particular, business software products, the 2008 report also highlights the need "to stop 'governmental use of unauthorised or illegal software'" noting that there had been a considerable progress in this area, with the APEC economies in 2006 agreeing that "central government agencies should use only legal software and other copyrighted materials and should implement effective policies intended to prevent copyright infringement on their computer systems and via the Internet" (USTR, 2008).

International collaborative efforts at training include:

- The United Nations Interregional Crime and Justice Research Institute (UNICRI) has, as part of its work programme, a project "aimed at enhancing the operational capacities of law enforcement agencies of selected States in the fight against ICT-related crime", which also covers issues related to counterfeiting and piracy, including digital piracy.
- The conduct, by WIPO, of regular workshops, training courses and conferences on intellectual property law enforcement, in line with requests from its member states.[4]
- The Intellectual Property Rights Experts Group established by the APEC Forum that co-ordinates work related to IPR. APEC also runs symposia and training activities in its member economies, such as the Workshop on IP Rights Enforcement in the Digital Era held in Vietnam in July 2007.
- In 2005, the International Chamber of Commerce launched the Business Action to Stop Counterfeiting and Piracy, which aims to connect the international business community. Its main activities include lobbying governments on issues relating to enforcement of IPR and training and information exchanges between its members.

Co-operation between rights holders

International examples of industry group alliances include the Business Software Alliance (BSA). The BSA is one of the most active industry trade groups in the area of copyright enforcement over digital products. It conducts its own studies, such as the 2006 Global Software Piracy Study, operates piracy hotlines around the world and actively and aggressively works to enforce copyright over software through litigation and other means including: a global Internet crawler system which identifies infringing files being offered on Internet sites; a USD 1 million rewards programme for those that report the sale of illegal software over the Internet; and liaising directly with governments. It has, for example, signed MOUs with national governments such as that of Qatar.

Further, in the United States there are various industry groups that actively seek to enforce copyrights of their various members whose products are being infringed online, mainly through litigation (though targets vary among groups). Some of the most prominent players include the MPAA, BSA, RIAA and ESA. There have been other attempts to create alliances between various organisations in an effort to combat digital piracy, such as the Computer Systems Policy Project in 2002, the Secure Digital Music Initiative in 1998, and the Copy Protection Technical Working Group in 1996.

Most of the co-operative schemes targeting digital piracy in the United Kingdom are between industry trade groups and government. The key groups are the Federation Against Copyright Theft (FACT), the British Phonographic Industry (BPI), the Education and Leisure Software Publishers Association (ELSPA), the Mechanical Copyright Protection Society (MCPS) and the Federation Against Software Theft (FAST). With regards to enforcement, many of these collaborations are *ad hoc* in nature, with industry groups and law enforcement bodies pooling intelligence or resources in particular cases as they arise. However, attempts to formalise links between such bodies are being made.

Examples of legal solutions to counter digital piracy include the French Olivennes Agreement and the UK Joint Memorandum of Understanding between music and film right holders and six major Internet service providers (see Box 2.4).

Government co-operation

Collaborative schemes aimed at the improved enforcement of copyright generally, or digital piracy specifically, may be considered in terms of *i)* international co-operation and *ii)* co-operation of government agencies within a particular economy.

International co-operation

The obvious example of international co-operation with respect to the enforcement of copyright is the work of dedicated enforcement agencies such as Interpol. With respect to the enforcement of copyright, the agency has set up the Interpol Intellectual Property Crime Action Group (IIPCAG), whose membership consists of representatives from national law enforcement and customs authorities, international intergovernmental organisations and cross-industry private sector representative bodies. Interpol and the IIPCAG combine efforts under the Interpol IP Crime Programme; one of the results of such efforts being the creation of the Database on International Intellectual Property Crime. Interpol has also signed Memoranda of Understanding (MOUs) targeted at co-operation of law enforcement (specifically dealing with piracy) with bodies such as the World Customs Organization (WCO).[5]

Other existing international organisations also have established non-binding processes for assisting in the enforcement of digital piracy. UNESCO, for example, has set up the Intergovernmental Copyright Committee that meets regularly to discuss copyright enforcement-related issues. The G8 has also endorsed the "Principles and Recommendations for Co-operative Investigation and Prosecution of Serious and Organized Intellectual Property Rights Crime" that set forth a foundation for future co-operation on criminal IP enforcement among G8 members (NIPLECC, 2008). The World Intellectual Property Organization (WIPO) also provides a forum for discussion of enforcement issues through its Advisory Committee on Enforcement, (ACE)[6] and has also set up an electronic forum – the IPEIS electronic forum – where members of the ACE can discuss matters concerning the enforcement of IP. An enforcement database and an IPR service centre have also been set up by the Asia-Pacific Economic Cooperation (APEC) Forum to aid in IPR enforcement and improve co-operation between member economies.

In the context of international co-operation with respect to copyright it is important to mention the June 2008 OECD Seoul Ministerial Declaration on the Future Internet Economy,[7] which includes two relevant statements:

- Combine efforts to combat digital piracy with innovative approaches which provide creators and rights holders with incentives to create and disseminate works in a manner that is beneficial to creators, users and our economies as a whole.
- Encourage new collaborative Internet-based models and social networks for the creation, distribution and use of digital content that fully recognise the rights of creators and the interests of users.

Local co-operation

Co-operation aimed at the enforcement of copyright is also evident within large economies. The United States, for example, has established the National Intellectual Property Law Enforcement Coordination Council (NIPLECC). The Council is "an interagency council responsible for co-ordinating US domestic and international IP enforcement activities"

(NIPLECC, 2008). As of 2005, it has been led by the newly created Office of the US Coordinator for International Intellectual Property Enforcement. The NIPLECC has initiated various inter-agency campaigns, the most prominent being STOP! (Strategy Targeting Organized Piracy, launched in October 2004). As a result of STOP!, there is an extensive and closely linked network within the United States which deals with enforcement of IPR.

With regards to digital piracy in particular, US government agencies such as the DOJ's Computer Crime and Intellectual Property Section (CCIPS) have launched schemes such as The Computer Crime Initiative to combat piracy in the digital sphere. Similarly, the National Intellectual Property Rights Coordination Center focuses on investigating organised IPR crime, and serves primarily as a 'liaison between private industry and law enforcement'. The Immigration and Customs Enforcement Cyber Crimes Centre, which works closely with the IPR Center, specifically targets piracy and related IP violations over the Internet.

Examples may also be found in countries like the United Kingdom and India. In the former, the National Intellectual Property Intelligence System has been set up by the UK Intellectual Property Office to "effectively join up Government, industry and enforcement agencies". Another UK initiative is the IP Crime Group (IPCG), created as part of the National IP Crime Strategy. The IPCG hosts regular meetings attended by representatives from bodies such as the Serious Organised Crime Agency and the Crown Prosecution Service. The IPCG also publishes an annual IP Crime Report, and has developed a national IP intelligence system database.

The Copyright Enforcement Advisory Council was established in India to strengthen the enforcement of copyright. Several Indian states and union territories have also set up either their own copyright enforcement cells or special cells in the Crime Branch to look after copyright offence cases, and "nodal officers" have been appointed to help co-ordinate activities and intelligence gathering.

Remedies for copyright infringement

Copyright is a private right. Thus, an infringement of copyright is an infringement of the copyright owner's private rights. The typical range of private right remedies (often called civil remedies) provided by courts – such as an order prohibiting further infringement (an injunction) and an order requiring payment of monetary compensation (an award of damages or an account of profits) – are available to copyright owners. In addition, however, some infringements of copyright are considered sufficiently heinous as to constitute an offence against the state. In that situation, the infringement will constitute a criminal offence, and courts may be empowered to exercise the more punitive sanction of imprisonment of the offender. Of particular relevance is the TRIPS Agreement, which requires all WTO Members to provide effective criminal remedies for commercial scale infringements, arguably regardless of whether there is a commercial (profit) purpose or intent.

International treaties

The WIPO Copyright Treaty and WIPO Performances and Phonograms Treaty simply require that countries ensure that enforcement procedures are available so as to "*permit effective action against any act of infringement*".[8] In contrast, the TRIPS Agreement contains quite detailed minimum standards for civil and administrative, as well as criminal, enforcement of copyright. No distinction is made for digital copyright infringement; TRIPS is technology-neutral in that its provisions apply regardless of the format or medium of the infringement. The TRIPS Agreement provisions on civil enforcement mandate the remedies typically provided by courts for infringement of private rights.[9] The provisions on administrative enforcement mandate special measures for seizure of infringing goods at borders.[10] The provisions on criminal enforcement require countries to provide criminal procedures and remedies "at least in cases of ... wilful copyright piracy on a commercial scale".[11]

National laws

Two types of remedies for copyright infringement tend to be distinguished at the economy level: *i)* civil remedies and *ii*) criminal penalties.

Civil remedies

In all jurisdictions, the standard civil remedies (such as injunctions, damages and/or account of profits) can be granted by a court for copyright infringement.[12] Various other civil remedies are often available. For example, in the United Kingdom, United States, France and New Zealand the copyright owner may also apply to the court to have the infringing articles delivered up to them.[13] In France, payment may be ordered of a certain sum of the amounts seized.[14] In Japan and New Zealand, the owner may demand "destruction of things which have constituted the act of infringement".[15] As a general rule, the civil enforcement processes available in national copyright laws provide comprehensive remedies for copyright infringement, including digital piracy.

Criminal penalties

In the United Kingdom, the showing in public (but not the copying) of a work, sound recording or film (qualified by the requisite knowledge) constitutes an offence.[16] In the United Kingdom, the United States, Australia and New Zealand (again, qualified by the requisite knowledge), any type of commercial dealing with copyrighted subject matter that constitutes infringement is also a criminal offence if it is "for commerce" or "in the course of business".[17] The Australian legislation also provides that conversion of copyrighted subject matter from hard copy or analogue form to "digital or other electronic machine-readable form" constitutes an aggravated offence.[18] In the United States, the relevant provisions specifically class making pre-release copyrighted subject matter "available on a computer network accessible to members of the public" to be a distribution of the type to which criminal penalties apply.[19]

Liability for circumvention of technological protection measures

Copyright law, in certain situations, extends liability beyond doing, contributing or authorising an act within the exclusive right of the copyright owner (OECD, 2005 and OECD, 2007). In response to the challenges of digital piracy, a particular type of extended liability has been introduced in recent times – namely, liability for circumventing, and facilitating the circumvention of, technological protection measures (TPM), and for removing or altering rights management information (together known as DRM, see Box 2.2) that seek to preclude the infringement of digital copyright material.

International treaties

The WIPO Copyright Treaty specifically creates obligations upon signatories to provide legal protection and remedies against the circumvention of technological protection measures and against alteration or removal of rights management information.[20] Similar provisions are included in the WIPO Performances and Phonograms Treaty.[21] The essence of these provisions is that member countries must grant "adequate legal protection and effective legal remedies" against the circumvention of technical protection measures and against the removal or alteration of rights management information, which the treaties as "information which identifies the work, the author of the work, the owner of any right in the work, or information about the terms and conditions of use of the work, and any numbers or codes that represent such information when any of these items of information is attached to a copy of a work or appears in connection with the communication of a work to the public".[22]

National laws

There is some variation in how national laws implement protection against circumvention of TPMs, but they generally produce similar results, even though the mechanisms for reaching these results may differ. The US provisions, for example, are very detailed. They are broad in scope, and preclude "descrambling", decrypting or acts that "otherwise avoid, bypass, remove, deactivate or impair a technological measure".[23] The prohibition applies irrespective of the intention or the knowledge of the person circumventing the TPM. The Australian provisions are very similar to the US provisions,[24] as a result of the Australia-US Free Trade Agreement of 2004. The United Kingdom, has implemented anti-circumvention protection in a somewhat different manner. Its provisions prohibit the manufacture for sale, or the hire, of devices "primarily designed" to circumvent technological measures, and provides specific remedies allowing search for, and seizure of, such devices.[25] All other European countries[26] have legislation prohibiting circumvention of TPMs, in accordance with the general requirements of the European Copyright Directive.[27] Yet another approach in seen in the Japanese legislation, which makes the circumvention of technological protective measure 'in bad faith' an infringement of copyright.[28]

Many national laws make circumvention of TPMs a criminal offence. The United States, for example, has specific criminal provisions that apply when there has been wilful breach of the prohibition for the purposes of "commercial advantage or private financial gain".[29] The French Intellectual Property Code includes criminal sanctions of up to six months imprisonment. The Australian legislation provides criminal sanctions of imprisonment of not more than five years.[30] Japan, too, has penal sanctions for persons who sell a circumvention device or provide a circumvention service.[31]

Notes

1. The following industry initiatives were presented in OECD (2007): Anti-Counterfeiting Group (AC-G), European Brands Association (AIM), ICC Business Action to stop Counterfeiting and Piracy (BASCAP), Global Anti-Counterfeiting Group (GACG), Global Business Leaders Alliance Against Counterfeiting (GBLAAC), International Anti-Counterfeiting Coalition (IACC), International Trademark Association (INTA), Association of European Trademark Owners (MARQUES).

2. See *www.webwisekids.org*.

3. See Indian Copyright Office website (*copyright.gov.in*) and Federation of Indian Chambers of Commerce and Industry website (*www.ficci.com*).

4. See WIPO Enforcement website: *www.wipo.int/enforcement/en/activities*.

5. See Interpol website on IP crime: *www.interpol.int/public/financialcrime/intellectualproperty*.

6. More information on the mandate and sessions of the ACE at: *www.wipo.int/enforcement/en/ace*.

7. See OECD (2008b).

8. World Intellectual Property Organization Copyright Treaty 1996 (hereafter WCT): art 14(2); World Intellectual Property Organization Performances and Phonograms Treaty 1996 (hereafter WPPT). art 23(2).

9. Agreement on Trade-Related Aspects of Intellectual Property Rights 1994, Annexe 1C to Agreement Establishing the World Trade Organization 1994 (hereafter TRIPS): Part III, Section 2.

10. TRIPS: Part III, Section 4.

11. TRIPS: art 61.

12. See, for example, Australian *Copyright Act* 1968 (hereafter AU): s115. See also the New Zealand *Copyright Act* 1994 (hereafter NZ): s120, US *Copyright Act* 1976 (hereafter US): §502. 504; UK Copyright, Designs and Patents Act 1998 (hereafter UK): s96; the Japanese Copyright Law 1970 (hereafter JP): arts 113(1)(ii), 114(1).

13. UK: s99(a); US §504; NZ: s122 (for civil proceedings), s132 (in criminal proceedings); the French Intellectual Property Code (hereafter FR): art L 335-7.

14. FR: arts L 332-1, 333-1.

15. JP: art 112(2); NZ: s134; *cf.* also Art. 46 of the TRIPS Agreement.

16. UK: s107(3).

17. AU: ss132AC-AJ; UK: s107; US: §506; NZ: s131.

18. AU: s132AK.

19. US §506(a)(1)(C).

20. WCT: arts 11 and 12.

21. WPPT: arts 18 and 19.

22. The treaties do not specify how such protection and remedies must be provided. This is left to individual countries to determine.

23. Digital Millennium Copyright Act (DMCA), s1201. The DMCA was enacted as part of the United States accession to the WIPO Copyright Treaty and is, therefore, an addition to the standard protections under the *Copyright Act 1976*.

24. AU: Part V, Division 2A.

25. UK: s296ZD (CRRR reg 24).

26. See, for example, the French *Droits d'auteur et les droits voisins dans la société de l'information* (DADVSI) arts 13 and 14; DE: s95a, s95c.

27. Directive 2001/29/EC of the European Parliament and of the Council of 22 May 2001.

28. JP: art 30(1)(ii).

29. US: §1203.

30. AU: Part V, Division 5, Subdivision E.

31. JP: art 120.

References

NIPLECC, (2008), "National Intellectual Property Law Enforcement Coordination Council", report to the President and Congress on Coordination of Intellectual Property Enforcement and Protection.

OECD (2005), "Digital Broadband Content: Music", DSTI/ICCP/IE(2004)13/FINAL, available at *www.oecd.org/sti/digitalcontent,* OECD, Paris.

OECD (2008a), *The Economic Impact of Counterfeiting and Piracy*, OECD, Paris.

OECD (2008b) "Shaping Policies for the Future of the Internet Economy", OECD Ministerial Meeting on the Future of the Internet Economy, available at: *www.oecd.org/futureinternet.*

Sports Report (2008), "Background Report on Digital Piracy in Sporting Events", Envisional Ltd and NetResults Ltd.

USTR (2008), Special 301 Report, Office of the United States Trade Representative, available at *www.ustr.gov*.

Youth Ambassador", (2006). "Youth Ambassador against Internet Piracy Scheme", press release, 22 November, available at *www.info.gov.hk/gia/general/200611/22/P200611220122.htm*, accessed 17 April 2008.

Chapter 4

Case Study: The Sports Rights Owners Sector

What is the sports rights owners sector?

The sale of rights to broadcasters is a major source of income for those sport organisations and leagues that control and own sports and sporting events. Depending on individual contracts, the sale of those broadcast and associated rights could include:

- live broadcasting of sporting events (TV and radio).
- live or delayed streaming of events on the Internet.
- delayed broadcasts/streaming of those sporting events.
- packaging of highlights.
- subsequent DVD releases as individual events or as part of sport compilations.
- subsequent rebroadcasts as “classic” or “historic” events.

Broadcast rights may include regional or geographical limitations; for example, some broadcast rights may only cover specific countries or regions, and broadcasters in individual countries would have to obtain the rights to broadcast in their own territories. The most obvious example of this is the Olympics, where the successful host broadcaster generally re-sells the rights to national broadcasters that want to provide coverage of the Olympic events in their countries.

Apart from the Olympics, the owners of sports rights include all of the world’s best known, and most high-profile sporting organisations, such as the *Fédération Internationale de Football Association* (FIFA, football), the *Fédération Internationale de l’Automobile* (FIA, motor racing), the National Basketball Association (NBA, basketball),Wimbledon (tennis), Cricket Australia and others.

While all of the portfolio of media rights listed above have value to the sports rights owners, the first two are clearly the most important and lucrative, as the live event is the product that has the greatest appeal to a prospective audience. While a short life span was identified in the main study as a special feature of digital products generally, the life span of live sports events is extremely short, and their value could deteriorate dramatically in minutes or even seconds (for example a horse race).

Once the sport event has ended, and the results known, interest (and therefore the value of the product) falls away sharply, even though its value does not totally disappear. Therefore, while acknowledging that there are a series of cascading commercial opportunities at risk from piracy involving sport broadcasting, this case study will largely focus on the first two elements in the list, representing the most valuable components of the broadcast rights, as well as being the segments that would be of greatest interest to pirates and their audiences.

What is at stake?

While this case study (and indeed its parent study) does not specifically measure the extent of digital piracy, it is nonetheless useful to briefly consider some examples of the magnitude of financial commitments around the world involving sport events that are at risk from piracy.

At an international level, broadcasters will pay USD 3.8 billion for exclusive rights to carry the 2010 Vancouver winter and 2012 London summer Olympics.[1] On a regional level, the Union of European Football Associations (UEFA) sold the rights to Euro 2008 for EUR 800 million,[2] while at a country level a consortium of Sony Television and World Sports Group was reported to have won a 10-year deal to telecast the Indian Cricket Board promoted Indian Premier League cricket matches for USD 1.026 billion.[3]

These are a few examples of the many sports broadcasting arrangements that exist around the world, and are intended to highlight the very significant investments made by sports broadcasters for the rights to those events. These investments, and in some cases possibly the future of individual sports, could be at significant risk if there is extensive piracy of the live broadcasts of events covered by those rights.

How is intellectual property created in this sector?

This case study specifically addresses the infringement of Intellectual Property Rights (as defined in the WTO TRIPS Agreement) when those infringements are associated with the sports owners rights sector. This means that while the sporting events themselves are not the subject of the study, as the main study on digital piracy makes clear, in most legal jurisdictions the broadcasting of the sporting events (and recordings made of those broadcasts) are protected by copyright law, and unauthorised transmission, streaming or recording of those broadcasts would infringe on those rights. It is these rights that are covered in this case study.

The infringement of those copyrights (in the terminology of this study "the piracy") would include the live streaming of the broadcasts as well as post-event downloads of full games or highlights.

How does piracy take place?

Earlier in this case study, it was noted that the original, live broadcast of sporting events is the time when those events have the greatest intrinsic value. Therefore, piracy that takes place during (or immediately after) the broadcast have both the greatest interest to consumers and the greatest potential of inflicting significant damage to the owners of the rights.

In this respect, as was the case with the examination of the audio/visual sector in Phase I of the overall study (OECD, 2008a), rapidly developing technology has provided would-be pirates with the means of intercepting and re-transmitting those broadcasts, in virtual real-time, thus enabling them to compete directly with the authorised broadcast or Internet streaming of the individual sporting events.

While it has been technically possible for many years to record TV broadcasts (recordable VHS tapes came into existence in the mid-1970s, and moved through DVDs and hard disks), these could not readily be re-transmitted to a broader audience without access to extensive and expensive broadcasting facilities, and piracy was restricted to serving the after-market, well after the live event had ended.

However, modern computers with cheap and easily fitted TV cards (increasingly found as standard fittings on home computers), can capture live TV signals from terrestrial, cable, satellite and Internet broadcasts, and with the use of some simple, freely available software can re-transmit those signals through the Internet. These captured signals can also be converted to digital data files and saved on a hard disk for subsequent viewing or transmission.

This capturing of the signal is the single most important pivot point in the piracy of sports broadcasts, as it is the means by which the pirates can – in virtual real time – illegally distribute the pirated content in direct competition with the rights owners, and at the precise time when the pirated content has the greatest value.

Software capable of producing content in a form that can be streamed to other Internet users is freely available, including in the form of a built-in capability in the popular Windows Media Player. In addition, the websites Sopcast and TV Ants (both based in China) have been identified[4] as two of the principal sites (amongst many that can be found on the Internet) that freely provide the software that permit users to generate and re-transmit a live broadcast stream at virtual real time within their own networks.[5]

The availability of this technology, and the increasing availability and affordability of Broadband connections, means that virtually anyone can now stream live programs through the Internet, which (like making music and video content available illicitly through the Internet) creates significant problems for the owners of broadcast and subsequent IP rights.

In essence, as the live transmission signals are captured by the computer's video card, they can be re-transmitted on the Internet either to individual recipients through Unicast transmissions, or as part of a transmission "swarm" using peer-to-peer (P2P) technology.

Unicast transmissions

The basic operation of Unicast transmissions on the Internet is that captured digital content (in this case live sport broadcasts, but could involve any TV transmission), is routed and/or stored on powerful servers and made available through direct Internet streaming from Unicast providers to individual viewers.

This one-on-one approach distinguishes Unicast transmissions from the swarm approach of peer-to-peer, which is discussed in the next section. Depending on the quality of the signal originally captured, the density (bit-rate) of the transmission and the width of the broadband connection, Unicast transmissions can offer transmissions of acceptable quality directly into any typical media player found in home computers. This means that viewers would be able to watch sporting events with little initial delay (caused by the need to create the initial signal buffer), or subsequent pausing to replenish the buffer (this may happen, for example, if there is a slowing down of the Internet signal).

In other words, these transmissions can provide virtual live feeds, in acceptable quality and with very little delay, in direct competition with the original, legitimate broadcast.

This one-on-one format, especially if fed to a large number of end users, can be very demanding of computer processing power and bandwidth, and there is generally a considerable cost associated with running such an operation. As a consequence many of these services either require a paid subscription, are supported by advertising or both. As such, they fit into that category of digital piracy (described in the principal study) that falls between the legitimate services offered by the broadcasting rights owner (whether a live TV signal and or live streaming on the Internet) and the generally free alternatives offered by the peer-to-peer networks.

There are clear incentives for sites offering Unicast services to appear as professional as possible in order to generate trust and attract customers, especially if subscription fees are charged. Many of these sites have sleek designs, smoothly operating interfaces and may feature logos from well know international companies, including sports rights owners and sponsors of the sport concerned. Additional apparent security (for the customer) is provided by the fact that payment for the transmission can be legitimately made through credit cards or PayPal. These are features that one would expect to find on legitimate site offering authorised transmissions (see Figure 4.1).

Such embellishments of the pirate sites encourage potential subscribers to view them as legitimate, and indeed in many cases it might be virtually impossible for most users to know (or even suspect) that such sites were not authorised to offer the streaming content.

If the quality of the streaming is good, then the pirate sites need not price their services at levels that are so far below those of legitimate sites that their provenance would be suspect. In any case, in these days of services being increasingly supported by advertising as well as subscription, price is not necessarily always a good basis for consumers to make judgements about the legitimacy of sites. Quite clearly this adds to the difficulties faced by the sports rights owners of the broadcast rights to stem the offerings from sites that can convince users into believing that they offer a legitimate service.

Figure 4.1. Screenshot of a typical Unicast site

Figure 4.2. Screenshot of MyP2P

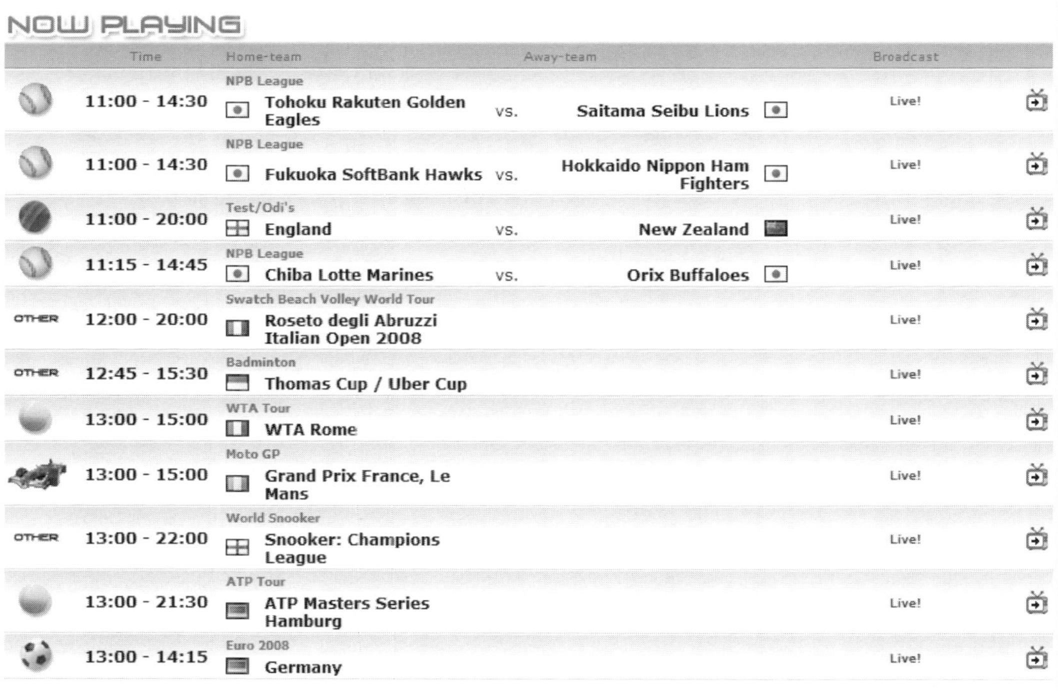

Figure 4.3. Sopcast screenshot

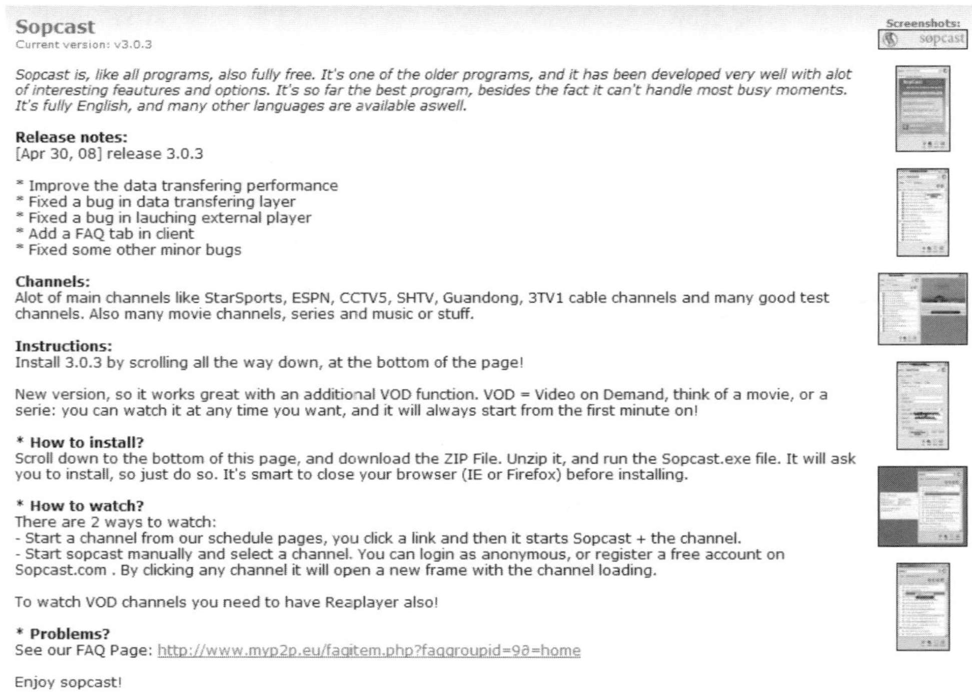

Sopcast
Current version: v3.0.3

Sopcast is, like all programs, also fully free. It's one of the older programs, and it has been developed very well with alot of interesting feautures and options. It's so far the best program, besides the fact it can't handle most busy moments. It's fully English, and many other languages are available aswell.

Release notes:
[Apr 30, 08] release 3.0.3

* Improve the data transfering performance
* Fixed a bug in data transfering layer
* Fixed a bug in lauching external player
* Add a FAQ tab in client
* Fixed some other minor bugs

Channels:
Alot of main channels like StarSports, ESPN, CCTV5, SHTV, Guandong, 3TV1 cable channels and many good test channels. Also many movie channels, series and music or stuff.

Instructions:
Install 3.0.3 by scrolling all the way down, at the bottom of the page!

New version, so it works great with an additional VOD function. VOD = Video on Demand, think of a movie, or a serie: you can watch it at any time you want, and it will always start from the first minute on!

*** How to install?**
Scroll down to the bottom of this page, and download the ZIP File. Unzip it, and run the Sopcast.exe file. It will ask you to install, so just do so. It's smart to close your browser (IE or Firefox) before installing.

*** How to watch?**
There are 2 ways to watch:
- Start a channel from our schedule pages, you click a link and then it starts Sopcast + the channel.
- Start sopcast manually and select a channel. You can login as anonymous, or register a free account on Sopcast.com . By clicking any channel it will open a new frame with the channel loading.

To watch VOD channels you need to have Reaplayer also!

*** Problems?**
See our FAQ Page: http://www.myp2p.eu/faqitem.php?faqgroupid=9∂=home

Enjoy sopcast!

Screenshots:
sopcast

Peer-to-peer (P2P) networks

As described in the principal study, as well as easier access to Broadband, the growing availability of increasingly sophisticated P2P software is a key driver that influences the demand and supply of pirated digital content, including live sports broadcasts.[6] In essence, P2P networks operate in the following way. Using one of the many available peer-to-peer networks, a person can establish a video stream that is available to others who join the stream. Software allows the stream to be broken up into small packets, which are then transmitted to others in the stream. As any given packet is downloaded and received by an end user, it then becomes immediately available to be uploaded to another user, thus creating a "swarm" on the network.

Generally, the more users are online the better the quality of the stream, with the end result that after some initial buffering of the signal, the quality of the transmission can be quite acceptable. It will, certainly be more than good enough for keen fans, who may not otherwise be able to access the live broadcast or are unwilling to pay for a legitimate service, and who will accept some degradation in quality in order to watch the sporting event live. A brief delay in the transmission (even of some minutes) would probably not significantly degrade the experience for many fans. In addition, the popularity of many sports, and the inability (or unwillingness) of many fans to watch the live transmission through legitimate services, would ensure that the on-line "community" for the more popular events would almost always have sufficient participants to ensure a successful P2P transmission.

The availability of web-based search tools means that it is not difficult to find P2P piracy streaming sites. However, for end users that task is made considerably simpler and more efficient through services that collate, index and promote sporting events. One such site is MyP2P,[7] which in a very lucid and comprehensive site (apparently supported by advertising from large and reputable multinational companies),[8] provides details of pirated live streaming covering a wide range of sports, from football (at the time of writing, all matches in the Euro '08 competition were being streamed) to cricket (which has a very large following in the sub-continent) – see Figure 4.2.

Using sites such as MYP2P simply require the downloading and installation of some free software, which then allows access to any of the live streams offered on that site. Detailed system information as well as on-line technical support is generally available to assist users in setting up their participation in the P2P networks (see Figure 4.3).

While the hosting site (sometimes called a tracker) organises and manages the different streams it does not actually distribute any video content; this is left to those users who participate in the peer-to-peer "swarm". In this respect, these live streaming sites operate in very similar fashion to other peer-to-peer and bit-torrent sites that are used to exchange and circulate digital content (even though not all of these activities involve the illicit exchange of copyright material).

Some distinctions between P2P and Unicast sites

While in the end the role of both of these forms of streaming technology, when applied to live broadcasts of sporting events, is to deliver a video stream to the end user, there are some factors that distinguish the forms of delivery:

- the Unicast sites, because of the need for significant computer processing power and Internet bandwidth to stream content to individual end users, generally apply subscription charges to end users wishing to access their services.
- while users of Unicast sites might have reason to believe that they are using a legitimate service provider (because of the professional appearance and a commercial feel to the sites), users of P2P sites would be under no illusion that they are using non-authorised sites to view the sporting events.
- users of Unicast sites are simply receivers of the video stream, but in the case of P2P sites they are (generally) simultaneously a receiver and a supplier of the content, and this may have different legal implications in some jurisdictions.

Other methods of pirating broadcasts of sports events

The Unicast and P2P methods are by far the most significant methods to pirate sports events, but they are not the only ones. Increasingly, sports events have been detected on user-generated content (UGC) sites; for example one sports rights owner reported that over 15 000 unauthorised clips were removed from one such site. In addition, while originally these sites were exclusively involved with archival content, *i.e.* recorded or delayed clips, in recent times a small number of UGC sites have been detected offering fans live streams via their servers.[9] While the majority of material on UGC sites probably does not infringe copyright, the growing popularity of these sites is likely to create further problems for sports rights owners in the future.

What are the market characteristics of the sports broadcasting sector?

In the broadest possible terms, the market for live sporting broadcasts would comprise all of the fans and other interested persons who are not present as spectators at the event, but who would nevertheless wish to share in the excitement of that event by watching a live broadcast/transmission.[10]

Many in this potential audience would be accommodated by the broadcast of the event either on free-to-air television, or on some form of paid television broadcast (perhaps delivered by satellite or cable) or (increasingly) by legitimate live streaming through the Internet.

However, this scenario leaves a proportion of potentially *legitimate* demand unfulfilled; that is customers who for one reason or another may be unable to access legitimate broadcasts/transmissions, for example:

- Legitimate live broadcasts may be region-specific and not available in all markets, perhaps because of contractual reasons, or because some markets may lack a sufficient audience base to make the broadcast financially attractive – cricket in continental Europe is a good example.
- Broadcast rights may exist, but only as delayed telecasts or through the provision of "highlights" packages, thus not meeting the requirement of potentially legitimate customers wishing to see the event "live".
- Large events with multiple activities (*e.g.* the Olympics or Wimbledon tennis) provide broadcasters with a choice of parallel sub-events that they can broadcast, which may leave a section of the potential audience unsatisfied with the broadcaster's choice; for example, a national TV channel may choose to focus on tennis matches featuring their home players, thus leaving other viewers unsatisfied.

This potentially legitimate market could consider itself as being forgotten or ignored by the broadcasters, and failing any other authorised alternative may turn to unauthorised live streams as a way of overcoming their "disadvantage". This attitude parallels the experiences in other digital content (such as popular TV shows or films) where licensing agreements and staggered roll-outs means that certain markets may have legitimate access to that content only after a considerable time from first release, or perhaps not at all, which encourages access to pirated alternatives.

In the live sports sector the effect of customer isolation from legitimate services is probably even more acute than other digital content, given that when the game has been played, and the result known, a lot of the interest wanes from the sporting event. This would act as a further encouragement for those fans to seek whatever alternatives are available to view the events live.

In addition to this unsatisfied and potentially legitimate market, there is also an *illegitimate market*, composed of those who, for a variety of reasons, will choose illicit ways of viewing those sporting events. If the general market for pirated content is any guide, then this market segment will be large, and will respond to the same incentives as those that drive demand for pirated music, films software and other digital content. That is, the market will be composed of persons ranging from those who simply cannot afford the cost of the legal options, through to those who would always choose the illicit alternatives (especially if they are free) regardless of the legitimate options that may be available to them.

In the context of live sporting events this would mean the use of unauthorised live transmissions or streaming, instead of paying the cost of legitimate services, perhaps available only from cable/satellite services on either period subscription or pay per view formats. Additionally, the illegitimate market would also include those who have generally permissive attitudes to digital piracy, and who see their participation in P2P swarms as a "community" activity, which from their perspective would enable others to "benefit" from their on-line participation in such peer-to-peer activities. These and other related drivers of digital piracy are discussed in more detail in the main study.

The combination of the potentially legitimate viewers who cannot access authorised services, and those who seek unauthorised streams even when legitimate services are available, describe the audience that may wish to access these alternative services. Many of those that have suitable Internet access, and have sufficient technical know-how to access Unicast, P2Pand even some UGC sites, may choose to use these to avail themselves of the illicitly provided live streaming or other access to those sporting events.

No effort has been made to estimate the potential size of this market, as this is outside the terms of reference for this study. However, as a pointer to the kind of audiences that might tune in to live sports broadcasts captured and streamed through the Internet, information provided to the OECD (Sports Report, 2008) noted that in December 2007 a total of 1.2 million views were registered on a SopCast channel that streamed a match from the US National Basketball Association (NBA). From a large sample of those connected to the stream, the largest proportion (around 78%), were found to be located in China. Similarly, the highest number of viewers ever recorded for a single P2P stream for cricket was more than 700 000. A number of other similar examples representing other sports were also reported.

In summary, even allowing for today's broadcasting and Internet technology, because of geography, contractual arrangements and commercial imperatives, it is unlikely that legitimate live broadcasts and Internet streaming will meet the needs of everyone on the planet, and because of this there will inevitably be a very significant potential audience whose needs will be unmet (or unmet at a price that they can afford or are willing to pay for), and many of these will turn to other services to view broadcasts of their preferred sporting events.

How are choices made in this market?

As described in the main study, classic economic theory postulates that, everything else being equal, demand will drift towards the lowest available price, and that the lower the price for any good or service the greater the demand is likely to be. Where the price of a desirable good is zero, or virtually zero, then demand can be expected to virtually unlimited.

Also described in the main study is the phenomenon that the digital piracy market, unlike every other sector that is subject to counterfeiting and piracy, not only has legitimate content providers at one price and illegitimate providers at a (generally) lower price, but also a category of content providers who are willing to provide (or share) content at zero or near-zero price. The understanding of this type of market calls for the extension of traditional economic models.[11]

In the sport rights case, the role of "lower cost" providers are largely played by the Unicast sites, which on the basis of either a general or pay per view subscription, will stream live events to subscribers. While these prices are likely to be lower than the legitimate services (who must charge full economic prices), as noted earlier the computer processing power and Internet bandwidth required to provide such services means that totally free Unicast streaming is unlikely, unless supported by third party advertising.

On the other hand, the use and content of P2P streaming sites are generally free to users, with their only obligation likely to be the joining of the peer-to-peer "swarm", that is, to act as uploaders as well as downloaders. The main reasons for this apparently irrational market behaviour (that is, the provision of content at zero price) are catalogued in the main study, but in essence these reflect reciprocity amongst participants, and other non-monetary drivers (such as social recognition within the on-line community). Indeed, the only "costs" that might be perceived by users of P2P sites would be the possibility of receiving some viruses or malware, or perhaps being subject to some action by the copyright owner.

The existence of these essentially free streaming services means that the sports rights owners, along with other rights owners whose content can be transmitted digitally, face competition of a kind not encountered during the extensive analysis of counterfeiting and piracy that result in physical goods (such as handbags, CDs and DVDs), which apart from the cost of production must then also be transported, distributed and sold.[12]

However, new economic paradigms or not, consumers generally tend to behave similarly, and the lure of streamed content at a low or even zero cost will inevitably attract customers, particularly where those consumers are unable or unwilling to access legitimate broadcasts or Internet streams. In keeping with normal market principles, all other things being equal, the greatest demand could be expected to occur at the lowest price offered, which with P2P streaming can be virtually zero.

Nevertheless, the fact that Unicast sites still exist means that price is not necessarily the only consideration, and again as described in detail in the main study, there are non-price factors that would drive demand. These could include legal concerns (a feeling of greater exposure in P2P streaming by virtue of being an uploader as well as a downloader), better quality (especially if P2P "swarms" are thin causing drop-outs and pauses for buffering) and a sense that paid sites are more professionally run and less likely to have malware embedded in their stream. The need for a certain degree of technical literacy in order to use the P2P networks, and perhaps even the belief by users that paid Unicast sites are legitimate, may also account for the continued existence of those sites.

It is also possible to conceptualise that similar factors (especially the issues of legality and quality) would be non-price factors that would encourage viewers to utilise legitimate broadcast and streaming services where these are available, rather than use the unauthorised alternatives.

The effects of these factors on the operation of the illegal sports broadcasting market can be seen in Figure 4.4, which shows the changing penetration over a three year period between paid Unicast sites (the falling trend line) and the generally free P2P sites that stream live broadcasts (the rising trend line).

Figure 4.4. Unicast *vs.* P2P streaming sites

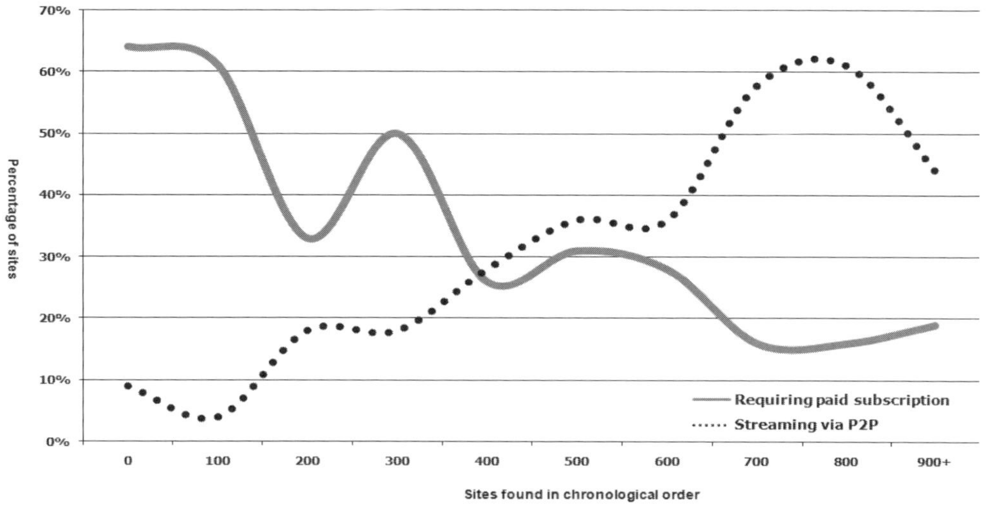

Source: Sports Report (2008), "Background Report on Digital Piracy in Sporting Events", Envisional Ltd and NetResults Ltd.

Industry responses

In a live sports broadcast context, rights holders have a number of possible responses open to them to deal with the distribution of pirated content:

- Take civil action against the owners of the site that is providing the pirated content (including taking down, closing or suspending the site).
- Take civil action against those who receive the pirated content.
- If possible, employ technological responses to curb the transmission of unauthorised streams or downloads.
- Where this is available, work with law enforcement authorities to apply criminal law provisions against providers and consumers of pirated content.
- Work with governments on regulations at appropriate points in the Internet delivery chains to facilitate the identification of, and action against, those involved in the delivery and consumption of unauthorised content.

These responses recognise that in the case of digital piracy, unlike other forms of counterfeiting and piracy, there are no producing factories to close down, goods in transit to intercept, or vendors to apprehend at the sale points. In the case of digital piracy the available responses are generally limited to attempting to close down offending sites (which experience has shown can be reopened much more quickly than physical production sites) and efforts to seek remedies through legal recourse where this is available to the rights holders.

This range of responses is also available to the sports rights owners, but because of the short life span during which live sporting events have maximum value (essentially the live and delayed broadcasts) those rights owners have very narrow windows of opportunity in which to respond to the rebroadcasting/streaming of pirated content.

Taking down offending sites

With respect to both Unicast and P2P sites, rights holders can screen sites in advance of major events to search for those sites offering the streaming of those events, and attempt to either prevent the streaming of the particular event or close the site down completely. Because of the relatively lengthy legal procedures involved (presuming that the sites are in jurisdictions where such legal remedies are available) this discovery needs to be well in advance of the event in question; discovery during the live event or after it is over is unlikely to be productive.

As an example, it was reported in July 2008 that China's National Copyright Administration (NCA) had warned more than 20 websites for illegally streaming Olympic programming.[13] The report also noted that some sites had ignored the warnings.

Some of the problems experienced by sports rights owners attempting to minimise the impact of Unicast and P2P sites have been:

- Difficulty in locating those sites sufficiently far in advance to take successful legal action.
- Many sites are located off-shore from where the sports event is being broadcast, making legal redress more complex, difficult and expensive.
- Sites often located off-shore, in jurisdictions with weak or poorly functioning legal systems or rules.
- Ability of sites to set up mirror sites (in different legal jurisdiction) to thwart the effect of any successful legal action.

- Ability and willingness of sites that are temporarily closed or even permanently taken down to set up new sites (even in the same legal jurisdiction) which require fresh legal action to be taken.

Because of the short "shelf-life" of live sports broadcasts, action against offending sites needs to be taken as early as possible, and ideally before the live broadcast starts. This requires extensive and costly monitoring of web sites to identify those who could be offering unauthorised streaming video of the broadcast. As shown earlier, many of these web sites are very well organised and provide lists of upcoming events days or weeks in advance of the live broadcast, and because they are generally located in difficult jurisdictions (from an IP perspective) they appear to be able to continue to operate in a very open manner.

Other Unicast and P2P sites that offer unauthorised "live" content can spring up at very short notice, which makes it very difficult for rights owners to respond quickly, and generally any action to take the sites down before the live broadcast is difficult, if not impossible.

Some sports rights owners take a longer term view, and attempt to work with the largest sites to minimise the appearance of their content. An example quoted by the sports rights sector (Sports Report, 2008) was the US Major League Baseball (MLB), which convinced SopCast to ban its content on that site. This was considered to be a positive development, but the discussions to achieve that outcome were extensive, and the MLB content immediately appeared on a rival site, thus largely negating any redress that the MLB may have gained from its initiative. However, this at least demonstrated that some negotiation was possible with the organisers of P2P sites, and more significantly that those organisers have a measure of control over the material that is streamed over their networks. However, success varies by organisation, and other sports right owners have not had such similar success in dealing with Sopcast.

Legal action against facilitators and sites

While sport rights holders have a range of available legal actions against sites and services that facilitate illegal content, the reported experience of the sector seems to indicate that even in IP friendly legal regimes such action is expensive, time consuming, by necessity largely reactive and not always effective. Sports rights owners reported that in a recent 12-month period four sporting organisations spent over EUR 1.3 million in attempts to challenge unauthorised streams of their content (Sports Report, 2008), highlighting the high cost of such action.

Essentially the problem in seeking legal redress (compensation, damages, etc. – as opposed to simply taking down the site) is that these can be very lengthy before those responsible (if found, and this can be difficult) can face an appropriate court. This allows sites to re-establish themselves in different jurisdictions, where fresh action would have to be restarted. In the view of the sports rights sector their experience highlights the difficulties of legal responses that are by necessity country-based, whereas the Internet operates in a global context, with few boundaries, either physical or legal, to hinder their operations.[14]

A number of specific examples were provided to the study team, involving sports as diverse as football (especially the English Premier League and UEFA), Australian football and cricket, some of which are discussed in the next section. In each instance, a favourable decision after lengthy court action proved ultimately ineffective, as new sites were quickly re-established to continue the unauthorised streaming of live sports events. In this sense, the experiences of the sports rights sector mirror the experiences of other rights holders whose content can be transmitted through the Internet; that is, action is difficult when the providers of unauthorised content are ephemeral, reside/operate in different legal jurisdictions and where consumers are very numerous (and also live in different jurisdictions) and are frequently active participants in the distribution of that unauthorised material. This is the nature of digital piracy, and these are the characteristics that make this sector different from traditional forms of counterfeiting and piracy involving physical products.

Technological responses

As well as legal recourse there are some technical responses that could minimise the streaming of unauthorised content. One technique is *geographical blocking* (commonly referred to as geo-blocking), which is frequently used on the Internet when content providers wish to restrict access to their content to specific geographical regions. An example of this is the BBC iPlayer, which limits access to those whose IP addresses are located in the United Kingdom.[15]

While such blocking technology would be useful to minimise (or at least localise) the streaming of unauthorised content, this requires the co-operation of the offending site in the first instance, and this is unlikely to be the case unless action is backed by some more direct pressure such as a court order, or perhaps (as in the case of the MLB mentioned earlier) an agreement between the rights owner and the web site concerned.

Filtering is also possible on a broader scale, for example by governments wishing to limit access to specific web sites by its citizens, but this enters into the very complex and sensitive field of personal freedom and rights, and apart from noting that this might be possible, it is not further explored.

Other technical responses, such as *traffic management and video fingerprinting* could also offer some scope for the sports rights sector to try to address the streaming of unauthorised content, but both have inherent problems. Traffic management is sometimes exercised by Internet service providers (ISPs) to manage their available bandwidth to ensure that it is not overwhelmed by large volume users (such as P2P streams). This traffic management tool is not really intended to block or slow down sites in response to possible IPR infringements, but could probably be technically feasible. Whether or not this may offer a further opportunity to prevent/ minimise the piracy of live sport broadcast is an issue for rights holders, ISPs and governments to address.

Video fingerprinting could be used to identify unauthorised copies, and apparently the technique is already being used in some music and video content, but the problem for the sports rights sector is that while it is technically feasible to track live video streams, there is no effective way of removing them.

In summary, while there are some nascent technical strategies that might possibly be used to minimise or prevent the illicit streaming of live sport events, these are not presently feasible, and would require considerable co-operation between governments, industry and consumers to deal with the sensitive issues associated with such responses. The recent involvement of ISP in some jurisdictions (for example the United Kingdom) to strengthen the opportunity of identifying and warning possible copyright infringers may indicate some movement in this direction.

Government regulation

From material provided to the OECD by the sports rights sector, there is a strong sense that Government regulations is seen as one of the best ways of ensuring protection for rights holders affected by digital piracy. There are two principal reasons for this perception. The first is that digital piracy is virtually a borderless activity, with the inherent problem that rights holders, providers and customers are likely to reside/operate in many different legal regimes, thus limiting the effectiveness of normal legal responses. In these circumstances, government regulation – especially if there were to be some kind of co-operation or consistency amongst governments in different legal jurisdictions – could be a way of facilitating responses and remedial action across those different jurisdictions.

Second, governments are often perceived as the only way of placing pressure on parties that have the potential ability to reduce the problem of digital piracy to act where they can, even if such action on their part is not seen as attractive or commercially desirable. In this context the Internet service providers are an obvious group for such attention, and their involvement in the battle against digital piracy (especially when this involves file sharing and P2P) streams) has been much reported. As this study was being written the announcement was made in the United Kingdom that ISPs had agreed on a plan with the music industry to address the problem of piracy. It was reported that under proposals by the UK government hard-core file sharers would be warned, and could see their broadband connections slowed.[16] A similar plan is currently under consideration in France.

No judgement is made here on whether this is an appropriate policy direction for governments and industry, but if this scheme does go ahead as planned (similar plans have been discussed in other jurisdictions) then this could also affect sports rights owners.

Specific industry examples

The sport rights sector made available to the OECD s a number of detailed examples of the experiences of specific sports in identifying and dealing with the piracy of live broadcasts (Sports Report, 2008 – from which much of the information in the following section was drawn). While the OECD was not in a position to verify the information provided in those examples, it has no reason to believe that they are not representative of the experiences of the sector. Only football, cricket and basketball are reflected here, principally because of the global popularity of these sports, their different characteristics and the range of data available for them. Other sports covered in the material made available to the OECD included baseball, tennis, Australian Rules football, rugby, motorsports, athletics, golf, American football and horseracing, thus indicating the very broad range of sports that are affected to a greater or lesser degree by unauthorised streaming of live events.

Football (soccer)

The experiences in football (soccer) have been reported by some of the largest leagues in Europe, including the Scottish and English Premier Leagues and Football League, Deutsche Fussball Liga (Germany) and Ligue de Football Professionnel (France), as well as football associations such as FIFA and UEFA.

Football is probably the most popular single sport in the world, and around 240 million people are reported to regularly play the game. FIFA estimated that the 2006 FIFA World Cup was shown on television in 214 countries across 376 channels with a cumulative television audience of 26 billion.[17] The magnitude of these numbers means that there is a very large pool of (frequently very passionate) supporters which would be a prime target for illicit streaming of live matches to meet the needs of those unwilling or unable to access legitimate broadcast and streaming services.

Monitoring of streaming sites involved activity across four major European football leagues, and the results are summarised in Table 4.1 below. The results for the Premier League and the German Bundesliga reflect season-long monitoring, while the Spanish La Liga, and the Italian Serie A comprised a snapshot of games across a single weekend.

Table 4.1. Sampling of illicit streaming sites affecting football

League	Monitoring period	Infringing sites	P2P-based	Unicast-based	Paid	Free	Viewers in China	Viewers outside China
English Premier League	2007-08 season	177	63%	37%	27%	73%	49%	51%
German Bundesliga	2007-08 season	85	96%	4%	10%	90%	73%	27%
Spanish La Liga	Snapshot during 2007-08 season	49	98%	2%	14%	86%	55%	45%
Italian Serie A		53	96%	4%	17%	83%	57%	43%
Average		*91*	*88%*	*12%*	*17%*	*83%*	*57%*	*43%*

This monitoring found that on average 91 infringing sites were found for each of the four leagues, although the Premier League appeared to face the worst problem, with 177 sites found by during season 2007-08 (barring the final weekend). The monitoring results indicated that a majority of all sites located were connected to P2P-based streaming, although the Premier League was subject to more attention from Unicast sites than the other leagues.

Significantly, the figures for infringing sites counted each located site once only; thus a portal site like MyP2P, which is likely to feature links to almost all football matches from the four leagues shown below was counted only once, despite providing links to many hundreds of individual matches. This of itself indicates the considerable number of streaming sites that were available to consumers, and correspondingly highlights the difficulties experienced by the rights owners to deal with them.

The sampled numbers also indicated that most viewers of monitored streams were based in China. While the distribution data was taken from only a small sample of P2P streams for each league, in the view of the monitoring company this reflected a large enough sample to show the large usage of such services in mainland China.

More specifically, the English Premier League found that during the 2007-08 season 177 different sites were located which contained or were connected to unauthorised streaming of matches. Of these, 122 (63%) used P2P methods to distribute the content, with the remaining 37% streaming direct through Unicast sites. Of the P2P streaming sites, 70 embedded streams into a web page, while 38 provided direct links to streams found on P2P streaming services or clients such as SopCast.

Compared to the other three leagues in Table 4.1, the Premier League appears to have a much greater problem with paid sites, where 27% of sites were accessible only after payment was made. This likely reflects the high level of demand for Premier League games and the realisation by owners of pirate sites that a section of viewers are prepared to pay for guaranteed access to Premier League games. In the context of the overall study these sites represent that segment of the market that exists between the legitimate broadcasters and those sites that provide (perhaps inferior services) at basically zero price.

As an indication of the impact of these illicit streaming sites, the site monitoring focussed its attention on the two largest illicit streams for an important match between the top two clubs in the Premier League in April 2008. The match was broadcast on the subscription channel Sky Sports in the United Kingdom and offered through various other broadcast means in over 200 other countries. Despite the availability of these authorised broadcasts, in total 238 000 viewers were estimated to have watched the game on the two SopCast streams.

A regional analysis of those viewers indicated that (consistent with the overall viewer location shown in Table 4.1) around 49% were located in China, with an additional 10% located in Hong Kong (China). The United Kingdom (home of the Premier League) comprised 13% of the viewers, with other developed economies also registering significant numbers of viewers, indicating that the use of illicit streaming services is widespread.

The experience of the other leagues indicates some differing characteristics to those of the Premier League. The most obvious difference is that the proportion of Unicast sites is considerably lower in the other leagues, so that P2P streams are by far the preferred mode of streaming those matches. As a natural consequence of this, there are commensurately fewer sites that require some form of payment, so that most access is through free sites. Finally, the proportion of viewers in China for the other leagues is significantly higher than for the Premier League, with the German Bundesliga being the highest, with 73% of all its illicit viewers being located in China. These differences reflect the respective popularity of the different leagues and their power to attract viewers, which in turn dictates to some extent the potential for service providers to seek payment from those viewers, perhaps by providing a higher quality and more reliable service.

While Unicast and P2P sites attracted the greatest number of viewers, post-event downloads of copies of entire matches are also available through the Internet, principally through file sharing networks such as bittorrent and eDonkey. The tracking of individual matches (in late 2007) between well known and popular teams such as Manchester United, Roma and Barcelona, attracted downloads in the tens of thousands, rather than the hundreds of thousands recorded for live streaming of matches. This is consistent with the characteristic of digital products that they have a comparatively short shelf life, and that live sporting events have the shortest effective life of all. Nevertheless even if smaller in number these illicit downloads would still affect the ability of the rights owners to maximise their returns from delayed telecast, packaged highlights and subsequent re-showing of some matches as "classics".

Cricket

While football has universal appeal, cricket is interesting because while its spread is more limited, it is enormously popular in South Asia (particularly India and Pakistan) as well as the United Kingdom, the West Indies, Australia, South Africa and New Zealand. The absence of significant fan bases (except for expatriates) in mainland Europe, the Americas, the bulk of Africa and China means that the target audience is quite different from that attracted to football. However, this did not prevent the collected data indicating that cricket had become the sport most affected by unauthorised streams of live events, particularly in India and Pakistan where the sport is immensely popular.

According to monitoring data, in early May 2008, a SopCast channel for an Indian Premier League game saw over 120,000 viewers watching the game. The highest number of viewers ever recorded for a single P2P stream for cricket was more than 700 000.

The monitoring of a number of major international cricket series (including the 2007 Cricket World Cup) located almost 950 cases which involved the unauthorised streaming of live cricket. Of these, around 260 were dedicated servers used by streaming sites to provide a Unicast stream direct to viewers. A further 230 cases were sites which required a subscription to access the cricket content, while P2P streams made up 280 cases.

The ratio of Unicast to P2P over the period is greater than has been the experience in other sports, but the collected data also shows that the proportion of peer-to-peer streaming used in cricket is growing. In 2005, most streams located for cricket were Unicast, for which the provider often required viewers to pay in order to cover the cost of dedicated servers with enough bandwidth to support the direct streaming to possibly thousands of simultaneous users. However, over the last two years P2P technology has become the dominant method used to stream live cricket through the Internet. Not only has the technology become faster and easier to use, but P2P-based streaming services are almost always free at the point of consumption because the costs of re-broadcasting streams are so much cheaper. This is because the technology generally makes use of viewer's own upstream bandwidth to deliver the content to those within the stream, and there is little requirement for fast servers.[18]

Data collected also showed that streams broadcast with the SopCast service make up the majority (around 60%) of P2P-based live streams for cricket.

As a further development in cricket, there has been a steadily growing number of embedded streams, or links to streams found on free hosting services, particularly those connected to free blogging hosting, (such as Blogspot). The technology is simple, and it takes only a few minutes to set up and post content of any kind to a Blogspot-hosted page. There is no cost and financial penalty if and when a page or site is removed. Relevant sub-domains (such as *ipl-on-sopcast.blogspot.com*) are freely available which help promote each new site and page. Users post comments which update others as to working streams in real time as games progress. This user behaviour is consistent with the "community" and co-operative peer approaches that were identified by the main study as drivers for the distribution and consumption of digital content (both licit and illicit).

Basketball

The data collected for basketball focused on the American National Basketball Association (NBA), which as a national series differs markedly from the international reach (in terms of participation) of both football and cricket.

While the NBA is a domestic series, its global popularity has continued to grow. Games from the NBA are currently broadcast in more than 200 countries and territories across 41 languages. NBA content is extremely popular in China, where the NBA has more than 50 authorised telecast partners. The popularity of the sport in China has been furthered by the successful introduction of a number of prominent Chinese-born players into the league, so that according to monitoring results, an estimated 78% of viewers of unauthorised streamed NBA games are located in China.

Over the past two NBA seasons, 172 sites have been found which provided unauthorised streams of NBA games online. Of these, just under three-quarters (74%) were related to P2P streaming and most of these were directly embedded streams, while around 30% took the route of offering links to available streams. SopCast is by far the most utilised P2P service found to offer authorised NBA streams (around 60%).

Audience sizes on these streaming sites are very large. On 16 December 2007, almost 1.2 million viewers were registered on a SopCast channel that streamed the *Dallas Mavericks* vs. *Houston Rockets* game. From a large sample of those connected to this game, by far the largest proportion of viewers (78%) was located in China. This is the largest documented viewing for a SopCast NBA stream and was likely enhanced by the inclusion of Yao Ming, a Chinese-born player who played in the game.

Such a result, involving as it did a domestic game of basketball, emphasises the global nature of the Internet, and the appeal of some sports that go well beyond their natural borders. Prior to the growth of broadband Internet access and the development of associated streaming technology, it would have been impossible for such a seepage to occur from legitimate audiences for live events, and emphasise the unique challenges being faced by the owners of rights associated with digital content (not just sports broadcasters) when dealing with technological developments that greatly facilitate digital piracy.

As well as its very large recorded audiences to streamed games, basketball is also interesting because of its efforts to create partnerships with some of the most widely viewed sites distributing unauthorised streams of NBA games. From those partnerships, services have been created with both PPLive and PPStream to allow a large number of NBA games to be offered live, and at no cost to the end user. Providing these legitimate streams of NBA programming may have had a positive effect as the average number of viewers of unauthorised streams of NBA games has declined since 2007, although the number of NBA games being streamed without authorisation has remained the same.

While strategic partnerships of this kind may offer an alternative way of dealing (at least in part) with unauthorised streaming of live matches, this model may not be available for other sporting leagues – particularly in the near-term when exclusive rights agreements have already been established for many sports in China and elsewhere.

Finally, the information provided with respect to the NBA notes that while China is the most significant problem area for the NBA, the problem of unauthorised streaming is not limited to there, and over the past year unauthorised streaming websites had also been identified in Europe, the United States, and Canada.

Conclusions

The purpose of carrying out a case study associated with the Phase II study on Digital Piracy is that it allows the detailed examination of the experiences of a particular sector affected by digital piracy to see how this actual experience reflects the analysis contained in the main study, as well as providing a practical snapshot of the experiences of a sector that distributes digital content in dealing with piracy.

The first point that becomes evident in the case study is that digital piracy can take a number of forms, and that each has its own particular characteristics. In the case of the sports rights owners sector, piracy takes the form of captured TV signals carrying live sports events that are then illicitly streamed through the Internet to customers around the world. Once captured, these events can also be made available subsequently by pirates as high-lights or clips on a delayed basis.

A characteristic of digital piracy identified in the main study is that digital content has a much shorter shelf life than other material that is counterfeited or pirated, and the case study has shown that live sporting broadcasts have the shortest shelf life of all. This is because unlike other digital content (such as films, music, software, books, etc) which can have appeal for some time after they are first released, the maximum value of the live sports broadcasts (and perhaps other live events such as concerts) is when the live event is happening. Once the events are over, and the result known, interest wanes very rapidly, although they still retain a value to owners as highlight or archive material.

As is the case with digital piracy generally, technological advances (ability to capture TV signals, high-speed broadband connections and streaming technology) has enabled pirates to easily and cheaply stream the broadcasts virtually in real time, and at a quality which although not perfect is sufficient to meets the needs of the target audience.

This real-time competition creates particular problems for the owners of the sport rights sector, as they essentially have a very narrow window of opportunity to respond to those illicit streams, and this both complicates their responses and makes them less effective; taking down a site once the sport contest is finished is little comfort to the aggrieved party. In addition, the cost of such action can be very high.

As far as the providers of the illicit streaming services are concerned, these reflect the characteristics identified in the main study; that is as well as a lower cost supplier of the streamed event (in this case Unicast sites which generally charge a lower price than legitimate services), there are also many content providers who provide content (through P2P sites) at zero, or virtually zero, cost to the final consumer. This is an economic model that does not exist in counterfeiting and piracy of physical goods, where there is always a cost of production, transport and distribution, and where goods are therefore never provided to consumers at zero price.

Although there are exceptions[19] generally the cost of illicit streaming is kept extremely low by virtue of the P2P protocols, which use the bandwidth available to the many participants in the P2P "swarms" to facilitate the streaming of the live events. Where streaming is carried out one a one-to-one basis (by Unicast sites) the cost of powerful servers and bandwidth is generally recouped through subscriptions, advertising or both.

The capabilities of computers and the Internet, as well as allowing real time competition for the original products, also allow the production and distribution of the streaming content from sites scattered around the world; including in unhelpful legal jurisdictions. Similarly, customers (including P2P participants) can be located virtually anywhere in the world, which means that as with other digital piracy the rights owners face a global problem, but have only local responses available to them.

In other words, legal action available to the rights owners may have to be taken at different locations from where the sporting event in question is taking place, and this can make action much more difficult and expensive and probably also less effective, as it is difficult to track down persons who operate the sites and take action against them.

Also consistent with other digital content (especially films and TV shows), complex commercial and licensing arrangements and lack of adequate customer bases means that legitimate broadcasts may not be available in all locations where there may be fans wishing to see particular sporting events, which means that otherwise potentially legitimate customers may look for other ways to view those events, thus creating more consumers to sustain the illicit streaming sites.

Further, the ability of pirates to compete with authorised broadcasts of live sporting event in virtual real time (attacking the content at the time that it has maximum value to the rights owners), and in virtually any market in the world, means that the owners of sport broadcasting rights face circumstance that are particularly difficult, and in the current international climate lack the means to adequately respond to these threats.

In conclusion, this case study has highlighted that the findings and conclusions of the main study on digital piracy, and that the issues identified for the consideration of policy makers and industry, are also relevant in their specific application to this sector.

Notes

1. Reported on the unofficial London Olympics website *www.the2012londonolympics.com*.
2. Reported at *www.theglobeandmail.com* on 19 June 2008.
3. Reported in *uk.biz.yahoo.com* on 20 June 2008.
4. See the background report on digital piracy of sporting events (Sports Report, 2008).
5. See *www.sopcast.org* and *www.TVAnts.com* (latter in Chinese although mirror sites available, such as *www.tvants_ppstream.com*).
6. It should be noted that P2P technology may be used for both legal and illegal purposes, and references in this annexe are intended to refer to its use as a tool to distribute pirated content. Many legitimate services are adopting P2P technology to provide access to content, with the authorisation of rights holders, and these services are not the subject of this chapter.
7. Can be found at *www.myp2p.eu*.
8. In the course of a few minutes of research in June and July 2008, advertising was seen from Hertz, EDF (French electricity utility), Primagaz, Promovacances (French holiday site) and the University of Phoenix.
9. Information provided by NetResult in correspondence to the OECD.
10. While there is also demand for delayed telecasts, highlights and other packaged content, these are not specifically considered in this brief analysis.
11. See the main study for a detailed discussion of this economic model and the drivers that affect the behaviour of suppliers of pirated digital products.
12. For further details see OECD (2008), *The Economic Impact of Counterfeiting and Piracy*.
13. "Websites warned for IPR infringement on Olympic streaming in China": *http://english.sina.com/sports/1/2008/0707/170268.html*.
14. Information provided in private correspondence to the OECD.
15. See *iplayerhelp.external.bbc.co.uk/help/about_iplayer/termscon*.
16. See *news.bbc.co.uk/2/hi/technology/7522334.stm*.
17. See *www.fifa.com/aboutfifa/marketingtv/factsfigures/tvdata.html*.
18. Though at least the initial uploader of the content must have access to a connection fast enough to stream the live broadcast at a speed which does not induce pauses or buffering.
19. For example, Bensports charges USD 100 per month (*www.bensports.tv*).

Annex A

Legal Aspects of Copyrights and Their Infringement

This annex presents legal foundations for copyrights and discusses regulations that refer to copyright infringements.

In simple terms, the doing of an act in relation to a work may be an infringement of copyright in the work if it is an act that is within the exclusive rights of the copyright owner of the work.[1] *This concept applies to digital piracy in the same way, and to the same extent, as it applies to any other act of copyright infringement.*

International treaties

The international treaties, in varying degrees, require that the owner of copyright in a work be provided with exclusive rights including the rights to reproduce the work and to make it available to the public in various ways.[2]

The Berne Convention,[3] for example, mandates that the exclusive rights of the owner of copyright in a literary and artistic work are the rights to reproduce, translate, adapt, publicly perform, ,and broadcast the work.[4] The TRIPS[5] Agreement also mandates these rights,[6] and introduces the further rights (for authors) of rental (in respect of computer programs and, in some circumstances, cinematographic works),[7] (for performers), fixation of unfixed performance and reproduction of such fixations(with respect to fixations on phonograms), (for producers) of phonograms direct or indirect reproduction of phonograms[8] and (for broadcasting organisations, or the owners of the copyrights in the subject matter of the broadcasts)[9] fixation, reproduction of fixation, re-broadcasting and communication to the public).[10] The WIPO Copyright Treaty[11] extends the rental right (to include works embodied in sound recordings),[12] grants a right of distribution of tangible copies and clarifies that the right of communication to the public also includes the making available of the work for interactive downloading, for example on the Internet.[13] The Rome Convention grants rights to performers to the fixation and live broadcast of their live performances and to the reproduction of such fixtures. It grants producers of phonograms rights to the direct and indirect reproduction of their phonograms and it grant broadcasting organisations rights to the fixation and simultaneous rebroadcast of their broadcasts and to reproduction of fixations. In addition, the Rome Convention contains an optional provision granting performers and producers of phonograms a right to remuneration of the broadcasting and other communication to the public of their commercial phonograms. Apart from the latter right, these rights are essentially also granted under the TRIPS agreement. The WIPO Performances and Phonograms Treaty[14] provide rights, similar to those in the Rome Convention and, as regards interactive downloading, similar to the WIPO Copyright Treaty in respect of sound recordings and performances fixed in sound recordings.[15]

National laws

The national laws of the major jurisdictions comply with the international treaties' provisions on the exclusive rights of the copyright owner.

Reproduction rights

The primary reproduction right is granted in all jurisdictions.[16] Copying includes digital reproduction; the UK legislation, for example, provides that copying includes 'storing the work in any medium by electronic means'.[17] Thus, digitisations of analogue copyright material, and reproductions of digital copyright material, are both acts within the exclusive rights of the copyright owner in all the major jurisdictions.

Distribution and communication rights

Various acts of distribution and communication of a copyright work are recognised as infringement of copyright in all OECD jurisdictions. Acts of distribution of tangible copies are usually limited to specific types of physical distribution. In the United Kingdom and New Zealand, for example, it is the act of "issuing copies of the work to the public" that constitutes infringement (and, in the case of sound recordings, films and computer programs, this includes any rental of copies to the public).[18] In Australia, the United Kingdom, New Zealand, and Japan, commercial dealings with infringing copyright subject matter – such as importing, distributing, selling and letting for hire – also constitute acts of infringement.[19] In the United States the commercial component is not necessary as the copyright holder has the exclusive rights to "distribute copies or phonorecords of the copyrighted work to the public by sale or other transfer of ownership, or by rental, lease, or lending."[20] Likewise, in Canada, the commercial component is also not necessary as secondary infringement covers any distribution of infringing material that would "affect prejudicially the owner of the copyright".[21]

In summary, the various tangible distribution rights in national laws cover the most common types of physical distribution of pirated copyright material.

Acts of intangible dissemination are often defined in broad terms, sometimes distinguishing between performance, broadcasting, communication to the public and the making available for interactive downloading on the Internet, the latter right frequently by using the language of the WIPO Copyright Treaty. Australia, for example, provides a general right to "communicate" to the public, which is defined to mean "make available online or electronically transmit".[22] In Japan the relevant act is termed "public transmission".[23] The United States, in contrast, provides protection against public communications and interactive downloading through a combination of the distribution, performance and display rights.[24] Although there is variation in the way in which national laws implement dissemination rights, there is no doubt that the acts of uploading, transmitting and downloading copyright material on the Internet are all within the exclusive rights of the copyright owner in all major jurisdictions.

Exceptions to copyright infringement

Not every act within the exclusive rights of the copyright owner constitutes an infringement. Both the international treaties and the national laws recognise various exceptions to the copyright infringement. If a particular act comes within one of these exceptions, there is no infringement of copyright. This concept applies to digital piracy in the same way, and to the same extent, as it applies to any other use of copyrighted content.

International treaties

Put in simple terms, the international treaties empower, but do not compel,[25] countries to provide exceptions to the rights of the copyright owner. That is to say, the treaties permit countries to make exceptions, but do not require them to do so. Furthermore, the general approach of the treaties since the Berne Convention is to describe the characteristics of the types of exceptions that are permitted, rather than to specifically list the exceptions that are permitted.

The Berne Convention does identify some particular activities (political speeches, legal proceedings, media reporting) in respect of which countries may provide an exception.[26] However, it also provides that countries may have other exceptions to the exclusive right of reproduction, so long as those exceptions satisfy the so-called "three-step test" – namely, that the acts permitted: are limited to certain special cases, do not conflict with the normal exploitation of the work, and do not unreasonably prejudice the legitimate interests of the copyright owner.[27] The TRIPS Agreement has adopted the Berne Convention three-step test for all copyright exceptions,[28] as has the WIPO Copyright Treaty and the WIPO Performances and Phonograms Treaty.[29]

National laws

The national laws of all countries contain a range of specific exceptions to copyright infringement. The main exceptions found in national legislation with relevance to digital piracy are: fair dealing or fair use; private or domestic use; incidental or temporary copies; and certain copying of computer programs. These exceptions are illustrated briefly in turn.

Fair dealing or fair use

Most jurisdictions allow for fair dealing or, in the case of the US 'fair use',[30] including: for the purpose of research or study,[31] and for the purpose of criticism, review or reporting the news.[32] In Australia, there is also a fair dealing provision for the purpose of parody or satire.[33] The German legislation provides exceptions for reporting the news as well as public speeches.[34] Similar provisions were made in the EU's Audio-Visual Media Services Directive that recommends that events of high interest to the public should be accessed on a fair, reasonable and non-discriminatory basis.[35]

Private and domestic use

Copying for "private and domestic" use is, to some extent, expressly exempted under the Australian, the United Kingdom and the New Zealand enactments. These incorporate acts labelled "time shifting" or "format shifting" in relation to subject matter such as sound recordings, films and broadcasts.[36] However, once a person deals with the subject matter in a commercial manner, such as selling, distributing or letting for hire, the exception no longer applies, nor does it apply if the source material is infringing.[37] In Australia, legislation provides that loan of the copy to a member of the lender's family or household for the member's private and domestic use does not constitute distribution.[38] In the United States, an infringement action may not be brought for the non-commercial use by a consumer of a digital or analogue audio recording device for making digital or analogue musical recordings.[39] In Japan, reproduction for private use is allowed with the additional requirement that compensation be collected from digital sound or visual recording equipment and media.[40]

Incidental or temporary copies

Exception provisions also exist for certain entities to make temporary copies in the course of communication,[41] or as part of a technical process of use,[42] or as incidental to another purpose.[43] In Australia, there is an express provision for the making of copies for the purpose of simulcasting (copying analogue subject matter for the purpose of simulcasting in digital form).[44] In Canada, a similar exception applies for a broadcaster copying a work into a format appropriate for broadcasting.[45]

Copying of computer programs

In Australia the copying of computer programs is allowed for certain purposes: reproduction for normal use or study, back-up copies, to make interoperable products, to correct errors, and for security testing.[46] A similar provision, while not as extensive, exists in the United Kingdom to cover observing, studying and testing of computer programs.[47] In the United States these exceptions are even more limited, and are only granted in the case of interoperability, maintenance or repair of computer hardware, or for archival purposes.[48] In Japan, reproduction is allowable for modification for computer programs,[49] while in New Zealand it can be made for back-up purposes.[50]

Notes

1. See, for example, Australian *Copyright Act* 1968 (hereafter AU): s36(1), s101(1). See also the New Zealand *Copyright Act* 1994 (hereafter NZ): s29(1), US *Copyright Act* 1976 (hereafter US): §501, the Canadian *Copyright Act* 1985 (hereafter CA): s27(1).
2. That is, to broadcast the works, to communicate the work to the public, for example through cable and to make the work available for interactive, on-demand communication, notably on the Internet.
3. *Berne Convention for the Protection of Literary and Artistic Works* (1886), amended on 28 September 1979 (hereafter Berne).
4. Berne: arts 9, 8 and 12, and arts 11, 11*bis* and 11*ter*, respectively.
5. *Agreement on Trade-Related Aspects of Intellectual Property Rights* 1994, Annex 1C to *Agreement Establishing the World Trade Organization 1994* (hereafter TRIPS).
6. TRIPS: art 9(1).
7. TRIPS: art 11.
8. TRIPS: art 14(1).
9. TRIPS: art 14(1).
10. TRIPS: art 14(3).
11. *World Intellectual Property Organization Copyright Treaty 1996* (hereafter WCT).
12. WCT: art 7.
13. WCT: arts 6 and 8, respectively.
14. *World Intellectual Property Organization Performances and Phonograms Treaty 1996* (hereafter WPPT).
15. WPPT: arts 13, 12, 14, 9, 8, 10 and 15.
16. For example: AU: s31(1)(a)(i), s85(1)(a) and s86(a); NZ: s30; the French *Intellectual Property Code* (hereafter FR): art 335-3; JP: art 21; US: §106(1).
17. See, for example, *Copyright, Designs and Patents Act 1998* (hereafter UK): s17(2).
18. UK: s18(1); NZ: s31.
19. AU: s37(1), s102(1), s38(1), s103(1); UK: s22, s23; NZ: s35, 36; the Japanese *Copyright Law 1970* (hereafter JP): art 113(1).

20. US: §106(3).

21. CA: s27(2).

22. AU: s31(1)(iv) and s10(1) (definition of "communicate"). For a detailed analysis of this right, see A. Christie and E. Dias (2005), "The New Right of Communication in Australia", *Sydney Law Review* 27, 237-262.

23. JP: art 23.

24. US: §§106(3), (4) and (5).

25. Some scholars argue that Berne does have one mandatory exception: quotations [Berne Art. 10(1)].

26. Berne: arts 2*bis*, and 10*bis*(1) and (2).

27. Berne: art 9(2).

28. TRIPS: art 13.

29. WCT: art 10; WPPT: art 16(2).

30. US: § 107.

31. AU: s40, s103C; UK: s29; NZ: s43.

32. AU: s41, s41A, s103A, s103B; UK: s30; NZ: s42.

33. AU: s41A, s103AA.

34. The German *Copyright Act 1965* (hereafter DE): arts 48-50.

35. Directive 2007/65/EC of the EC Parliament and of the Council of 11 December 2007: Art. 3k.

36. AU: s109A (with respect to sound recordings), s110AA (with respect to cinematograph films), s111 (with respect to broadcasts); UK: s70 (with respect to broadcast and cable programmes); NZ: s84 (with respect to time shifting).

37. AU: s109A(3), s110AA(3), s111(3); UK's *Copyright and Related Rights Regulations 1996* (hereafter CRRR), reg 19(2).

38. AU: s109A(4), s110AA(4), s111(4).

39. US: §119. US Code, Title 17, §1008.

40. JP: art 30.

41. AU: s43A, s111A; US: §112; EU: art 5.

42. AU: s43B, s111B; UK: s28A (as amended by CRRR reg 8(1)); EU: art 5.

43. See, for example, UK: s68 (with respect to incidental recording for purposes of broadcast or cable programme); NZ: s41, s85.

44. AU: s47A, s111A.

45. CA: s30.9.

46. AU: ss47B-47F.

47. UK: s50BA (introduced by CRRR reg 15).

48. US: §117.

49. JP: art 47 *bis*.

50. NZ: s80.

Annex B

OECD Council Recommendation on Broadband Development (2004)

The OECD Council recommends that, in establishing or reviewing their policies to assist the development of broadband markets, promote efficient and innovative supply arrangements and encourage effective use of broadband services, Member countries should implement:

- Effective competition and continued liberalisation in infrastructure, network services and applications in the face of convergence across different technological platforms that supply broadband services and maintain transparent, non-discriminatory market policies.
- Policies that encourage investment in new technological infrastructure, content and applications in order to ensure wide take-up.
- Technologically neutral policy and regulation among competing and developing technologies to encourage interoperability, innovation and expand choice, taking into consideration that convergence of platforms and services requires the reassessment and consistency of regulatory frameworks.
- Recognition of the primary role of the private sector in the expansion of coverage and the use of broadband, with complementary government initiatives that take care not to distort the market.
- A culture of security to enhance trust in the use of ICT by business and consumers, effective enforcement of privacy and consumer protection, and more generally, strengthened cross-border co-operation between all stakeholders to reach these goals.
- Both supply-based approaches to encourage infrastructure, content, and service provision and demand-based approaches, such as demand aggregation in sparsely populated areas, as a virtuous cycle to promote take-up and effective use of broadband services.

- Policies that promote access on fair terms and at competitive prices to all communities, irrespective of location, in order to realise the full benefits of broadband services.
- Assessment of the market-driven availability and diffusion of broadband services in order to determine whether government initiatives are appropriate and how they should be structured.
- Regulatory frameworks that balance the interests of suppliers and users, in areas such as the protection of intellectual property rights, and digital rights management without disadvantaging innovative e-business models.
- Encouragement of research and development in the field of ICT for the development of broadband and enhancement of its economic, social and cultural effectiveness.

The Council also instructs the Committee for Information, Computer and Communications Policy to monitor the development of broadband in the context of this Recommendation within three years of its adoption and regularly thereafter.

Annex C

The Seoul Declaration for the Future of the Internet Economy

WE, the Ministers and representatives of Australia, Austria, Belgium, Canada, Chile, the Czech Republic, Denmark, Egypt, Estonia, Finland, France, Germany, Greece, Hungary, Iceland, India, Indonesia, Ireland, Israel, Italy, Japan, Korea, Latvia, Luxembourg, Mexico, the Netherlands, New Zealand, Norway, Poland, Portugal, Senegal, the Slovak Republic, Slovenia, Spain, Sweden, Switzerland, Turkey, the United Kingdom, the United States of America and the European Community, assembled in Seoul, Korea, on 17 and 18 June 2008 to discuss the future of the Internet Economy.

WE STATE our common desire to promote the Internet Economy and stimulate sustainable economic growth and prosperity by means of policy and regulatory environments that support innovation, investment, and competition in the information and communications technology (ICT) sector. We will work with the private sector, civil society and the Internet community to secure the ICT networks that underpin the Internet Economy as well as to take measures to protect the users of the Internet Economy, including the necessary cross-border co-operation.

WE ARE DETERMINED to work together to promote ubiquitous access to ICT networks and services enabling widespread participation in the Internet Economy. The further expansion of the Internet Economy will bolster the free flow of information, freedom of expression, and protection of individual liberties, as critical components of a democratic society and cultural diversity. We will also work to use the tools of the Internet Economy to address global challenges, such as climate change. In moving forward, we recognise the significant foundation that the 1998 *OECD Ministerial Conference on electronic commerce* provided to the nascent Internet Economy and take note of the outcomes of the 2003 and 2005 *World Summit on the Information Society* (WSIS).

WE SHARE a vision that the Internet Economy, which covers the full range of our economic, social and cultural activities supported by the Internet and related information and communications technologies (ICT), will strengthen our capacity to improve the quality of life for all our citizens by:

- Providing new opportunities for employment, productivity, education, health and public services as well as addressing environmental and demographic concerns.
- Acting as a key driver for the creation of enterprises and communities and stimulating closer global co-operation.
- Enabling new forms of civic engagement and participation that promote diversity of opinions and enhance transparency, accountability, privacy and trust.
- Empowering consumers and users in online transactions and exchanges.
- Reinforcing a culture of security which applies to information systems and networks, and their users.
- Developing an increasingly important platform for research, international science co-operation, creativity and innovation in many different sectors.
- Creating opportunities for new economic and social activities, applications and services through ubiquitous and seamless access to communication and information networks.
- Promoting a global information society based on fast, secure and ubiquitous networks which connect billions of people, machines and objects.

WE AGREE that our challenges are, through an appropriate balance of laws, policies, self-regulation, and consumer empowerment, to:

- Expand Internet access and use worldwide.
- Promote Internet-based innovation, competition, and user choice.
- Secure critical information infrastructures, and respond to new threats.
- Ensure the protection of personal information in the online environment.
- Ensure respect for intellectual property rights.
- Ensure a trusted Internet-based environment which offers protection to individuals, especially minors and other vulnerable groups.
- Promote the secure and responsible use of the Internet that respects international social and ethical norms and that increases transparency and accountability.

- Create a market-friendly environment for convergence that encourages infrastructure investment, higher levels of connectivity and innovative services and applications.

WE DECLARE that, to contribute to the development of the Internet Economy, we will:

1. Facilitate the convergence of digital networks, devices, applications and services, through policies that:

- Establish a regulatory environment that assures a level playing field for competition.
- Uphold the open, decentralised and dynamic nature of the Internet and the development of technical standards that enable its ongoing expansion and contribute to innovation, interoperability, participation and ease of access.
- Stimulate investment and competition in the development of high capacity information and communication infrastructures and the delivery of Internet-enabled services within and across borders.
- Ensure that broadband networks and services are developed to attain the greatest practical national coverage and use.
- Encourage a more efficient use of the radio frequency spectrum to facilitate access to the Internet and the introduction of new and innovative services, while taking into account public interest objectives.
- Encourage the adoption of the new version of the Internet protocol (IPv6), in particular through its timely adoption by governments as well as large private sector users of IPv4 addresses, in view of the ongoing IPv4 depletion.
- Ensure that convergence benefits consumers and businesses, providing them choices with respect to connectivity, access and use of Internet applications, terminal devices and content, as well as clear and accurate information about the quality and costs of services.

2. Foster creativity in the development, use and application of the Internet, through policies that:

- Maintain an open environment that supports the free flow of information, research, innovation, entrepreneurship and business transformation.
- Make public sector information and content, including scientific data, and works of cultural heritage more widely accessible in digital format.
- Encourage basic and applied research on the Internet and related ICTs.
- Encourage universities, governments, public research, users and business to work together in collaborative innovation networks and to make use of shared experimental Internet facilities.
- Combine efforts to combat digital piracy with innovative approaches which provide creators and rights holders with incentives to create and disseminate works in a manner that is beneficial to creators, users and our economies as a whole.
- Encourage new collaborative Internet-based models and social networks for the creation, distribution and use of digital content that fully recognise the rights of creators and the interests of users.
- Strengthen the development of human resources to take full advantage of the Internet and related ICTs, and further develop ICT skills and digital and media literacy.

3. Strengthen confidence and security, through policies that:

- Protect critical information infrastructures at national and international levels from security risks.
- Strengthen the resilience and security of the Internet and related networked ICT systems and devices to meet the increasing demands and needs of our economies and societies.
- Reduce malicious activity online through reinforced national and international co-operation among all stakeholder communities in their steps for effective prevention, protection, information sharing, response, business continuity and recovery.
- Ensure the protection of digital identities and personal data as well as and the privacy of individuals online.

- Ensure that consumers benefit from effective consumer protection regimes and from meaningful access to fair, easy-to-use, and effective dispute resolution mechanisms, including appropriate redress for economic harm resulting from online transactions.
- Encourage collaboration between governments, the private sector, civil society and the Internet technical community in building an understanding of the impact of the Internet on minors in order to enhance their protection and support when using the Internet.
- Promote research to address emerging security threats.

4. Ensure that the Internet Economy is truly global, through policies that:

- Support expanded access to the Internet and related ICTs, especially for people in developing countries.
- Recognise the potential of the Internet and related technologies to provide enhanced services to people with disabilities and special needs.
- Recognise the importance of a competitive environment for the successful growth of the Internet Economy and the opportunities this can bring for development, particularly for people and regions with the most limited economic means.
- Promote use of Internet and related ICT networks by all communities as well as the creation of local content and multi-language translations to improve economic and social inclusion of people with different capabilities, education, and skills, and to preserve cultural and linguistic diversity.
- Facilitate the introduction of internationalised domain names (IDNs) while ensuring the integrity and stability of the Internet.
- Increase cross-border co-operation of governments and enforcement authorities in the areas of improving cyber-security, combating spam, as well as protecting privacy, consumers and minors.
- Harness the potential of the Internet to tackle global challenges such as improving energy efficiency and addressing climate change.

WE WELCOME the OECD report *Shaping Policies for the Future of the Internet Economy,* **RECOGNISE** its importance and **COMMEND** its consideration by OECD Member countries and non-member economies in developing their policies to support the Internet Economy.

WE COMMIT to working collectively with all stakeholders towards implementing and reviewing, as appropriate, the understanding that we have achieved in this Declaration in order to maintain its relevance to future challenges and opportunities confronting our economies and societies.

WE INVITE the OECD to further the objectives set out in this Declaration, through multi-stakeholder co-operation, by:

- Analysing the future development of the Internet Economy, namely: *i)* the important role and contribution of the Internet and related ICTs as a driver of innovation, productivity and economic growth; *ii)* the economic, social and cultural impacts of emerging Internet technologies, applications and services, including virtual worlds, sensor-based networks and social networking platforms.
- Based on this analysis, developing and promoting policy and regulatory principles, guidelines, other instruments and best practices for the future development of the Internet Economy.
- Researching the impacts of Internet and related ICTs in addressing climate change and improving energy efficiency.
- Examining the role of various actors, including intermediaries, in meeting policy goals for the Internet Economy in areas such as combating threats to the security and stability of the Internet, enabling cross-border exchange, and broadening access to information.
- Improving statistical systems to measure the changing access and use of the Internet and related ICT networks by citizens, businesses and institutions in order to provide reliable measures of evolving uses and the impact of the Internet on economic performance and social well-being.
- Assessing the application of current OECD instruments addressing consumer protection and empowerment, privacy and security in light of changing technologies, markets and user behaviour and the growing importance of digital identities.
- Recommending the development of OECD instruments that provide guidance in the formulation of policies for the development and use of converged communication networks.

- Continuing multidisciplinary work looking at the challenges and good practices of e-government and public sector transformation.
- Supporting measures and mechanisms to implement more effective cross-border co-operation.
- Conveying this Declaration and the OECD report "Shaping Policies for the Future of the Internet Economy" to all relevant international bodies and organisations, including G8, the ITU, WIPO, and UNESCO.
- Reinforcing co-operative relationships and mutually beneficial collaboration with the Asia-Pacific Economic Co-operation, the Council of Europe as well as the Internet technical community, the private sector and civil society within fora such as the Internet Governance Forum.
- Reviewing within three years of its adoption, and thereafter as appropriate, the progress made at national and international levels in light of this Declaration.

OECD PUBLISHING, 2, rue André-Pascal, 75775 PARIS CEDEX 16
PRINTED IN FRANCE
(93 2009 06 1 P) ISBN 978-92-64-06450-8 – No. 56883 2009